Climate Change and Biodiversity

Climate Change and Biodiversity

— *Editors* —
Nanda Nautiyal
Vir Singh
Neelendra K Joshi
Akanksha Rastogi

BIOTECH
2012
BIOTECH BOOKS®

ISBN 978-81-7622-254-9

Published by: **BIOTECH BOOKS®**
4762-63/23, Ansari Road, Darya Ganj,
New Delhi - 110 002
Phone: +91-011-23262132
E-mail: biotechbooks@yahoo.co.in

Printed at: **Chawla Offset Printers**
Delhi - 110 052

PRINTED IN INDIA

Every form of life is unique, warranting respect, regardless of its present worth to man.

–Henryk Skolimowski

All is One

The diversity of the world is a fact,

But also its unity.

He who does not see the unity

Lives amidst isolated facts.

Who has grasped the unity

Lives with fullness of life.

The wisdom of life means

Comprehending the unity of the Cosmos.

–Henryk Skolimowski

— *Dedicated to* —

Professor Henryk Skolimowski, the Father of Eco-philosophy, who teaches and practices life-enhancing philosophy—an imperative of the destiny of our living planet

Preface

Biodiversity is the crux of life.
Biodiversity in sugar, in chilly-
Biodiversity in the spice of life.
Biodiversity is at the heart of Evolution of Life.
Biodiversity is what life wishes to go through—
Variety of life, varieties of life, galaxy of life on Earth.
Biodiversity is what life blossoms with—
Ecosystems, countless species, innumerable genes.
Biodiversity is the meaning of life.
Biodiversity is the empowerment of life.
Biodiversity is the language of natural evolution.
Biodiversity is foundation of development.
Biodiversity infuses resilience in life.
Biodiversity is the mantra of sustainability.
Biodiversity combats entropy.

Biodiversity strikes a balance.
Biosphere reverberates with biodiversity,
Biodiversity puts the cosmos in balance.
We humans are but a single thread
In the complex web of life.
Biodiversity is the greatest wonder on Earth.
Biodiversity is oneness of life.
Biodiversity is philosophy of life.
Biodiversity is the beauty of life.
Biodiversity is glory of life.
Biodiversity is the song of life.

– Vir Singh

Biodiversity, in essence, represents all colours of life. Disappearance of even a single colour of life invites elimination of whole life. Evolution of biodiversity on Earth was the climax of life on Earth. The evolution never comes to an end. Evolution always keeps the things evolving. Climax too is not the 'final stage'. The climax too is continuously evolving into higher levels of climax, which are more beautiful, more aesthetic, more sustainable. Higher are the orders of biodiversity, more intensive will be the beauty of nature and of cosmos. More beautifully a thing exists, more sustainable it will be. Climax of biodiversity, beauty and sustainability go hand in hand.

The on-going change in climate order is influencing life in one way or the other. It poses a severe threat to the very crux of life on Earth—the biodiversity. The reasons of climate change leading to global warming and unusual weather behaviour could be several, including cosmic factors, but anthropogenic factors appear to be dominant. Accelerated pace of species' extinction due to anthropogenic activities revolving round development intervention was clearly noticeable much earlier than the indicators of climate change came to the fore more conspicuously. Climate change has further added fuel to fire. Biodiversity erosion and extinction of species and/ or their genotypes have further aggravated the problems and added to miseries of the planet.

Accelerated glacier melt, increasing sea level, reduced flow of water in rivers and streams in lean period, longer period of drought spell, unprecedented floods during monsoon season, increased rates of soil erosion, devastating landslides, inundation of large areas, unbearable summer season, etc. are the striking indicators of climate change. Erosion of biodiversity due to a mix of direct and indirect environmental factors is undoubtedly the greatest threat to life on earth.

This volume attempts to collate a number of articles written by eminent authors which provide ample information about climate change impact on biodiversity. Intricate relationships amongst biodiversity, biosphere and biological bloom – a relationship that underlines the natural path of evolution – have been discussed at length. Divided in three parts – *viz., Climate Change: Risks and Vulnerability to the Biodiversity, Conservation of Biodiversity: New Perspectives and Applications, and Biodiversity Restoration: Innovative Management approaches* – the book presents many facets of biodiversity against the backdrop of climate change. The views expressed, however, are of the authors and need not corroborate with those of editors or their organizations and of the publisher.

We hope that the book will be highly useful in generating interest among conscious people for undertaking vital task of protecting, conserving, and enhancing biodiversity while doing everything possible contributing to climate change mitigation. The editors would like to appreciate the efforts of all those committed to the cause of biodiversity, life and Mother Earth. Comments and suggestions of the readers and users of the book will be highly valuable.

Vir Singh

Nanda Nautiyal

Akanksha Rastogi

Neelendra K Joshi

Contents

Part II
Biodiversity Conservation: New Perspectives and Applications

Part III
Biodiversity Restoration: Innovative Management Approaches

List of Contributors

Arora, Sucheta. Deptt of Agril Communication, College of Agriculture, G.B. Pant University of Agriculture and Technology, Pantnagar, U.S. Nagar, Uttarakhand- 263145

Arya, Dipti. Department of Genetics and Plant Breeding, College of Agriculture, G.B. Pant University of Agriculture and Technology, Pantnagar, U.S. Nagar, Uttarakhand- 263145

Bari, MS. Associate Professor, Department of Agroforestry, Hajee Mohammad Danesh Science and Technology University (HSTU), Dinajpur 5200, Bangladesh;
e-: shafiqul_bari@yahoo.com

Bashir, Altamash. Post Graduate in Forestry, Sher-E-Kashmir, University of Agricultural Sciences and Technology of Kashmir, Shalimar, Srinagar, India- 191121;
e-: altamashwani@rediffmail.com

Chandola, Pawanika. Department of Plant Physiology, College of Basic Sciences and Humanities, G.B. Pant University of Agriculture and Technology Pantnagar, Uttarakhand – 263145, India

Chauhan, Asheesh. Deptt of Agril Communication, College of Agriculture, G. B. Pant University of Agriculture and Technology, Pantnagar, U.S. Nagar, Uttarakhand- 263145

Chauhan, R.S. College of Fisheries. GB Pant University of Agriculture and Technology, Pantnagar – 263145, Uttarakhand, India

Gusain, Poonam. Department of Microbiology, CBSH, GB Pant University of Agriculture and Technology, Pantnagar – 263145, Uttarakhand, India; e-: poonamgsn@gmail.com

Joshi, Bhasker. R.H. Govt. P.G. College, Kashipur (Kumaun University, Nainital) Uttarakhand; e-: bhaskerjoshiphd@yahoo.com, bhaskerjoshiphd@in.com

Joshi, Bhawana. Department of Plant Physiology, College of Basic Sciences and Humanities, G.B. Pant University of Agriculture and Technology Pantnagar, Uttarakhand–263145, India

Joshi, Neelendra K. Department of Entomology, 501, ASI Building, The Pennsylvania State University, University Park, 16802, PA, USA

Jyoti. Department of Plant Physiology, College of Basic Sciences and Humanities, G.B. Pant University of Agriculture and Technology, Pantnagar-263145, Uttarakhand, India

Kar, CS. Department of Genetics and Plant Breeding, College of Agriculture, G.B. Pant University of Agriculture and Technology, Pantnagar, U.S. Nagar, Uttarakhand- 263145

Khan, MA. Faculty of Forestry, Post Graduate in Forestry, Sher-E-Kashmir, University of Agricultural Sciences and Technology of Kashmir, Shalimar, Srinagar, India-191121

Kumar, Prem. KVK, SKUAST-J, Jammu, India

Kumar, Rajesh. Department of Microbiology, CBSH, GB Pant University of Agriculture and Technology, Pantnagar – 263145, Uttarakhand, India

Kumar, Rajesh. College of Fisheries. GB Pant University of Agriculture and Technology, Pantnagar – 263145, Uttarakhand, India

Limbu, Nandita. Department of Genetics and Plant Breeding, College of Agriculture, G.B. Pant University of Agriculture and Technology, Pantnagar, U.S. Nagar, Uttarakhand- 263145

Majeed, Mumtaz. Post Graduate in Forestry, Sher-E-Kashmir, University of Agricultural Sciences and Technology of Kashmir, Shalimar, Srinagar, India- 191121;
e-: mumtazsofi@rediffmail.com, sofimumtaz_sk@yahoo.co.in

Nag, Amrita. SR Luthra Institute of Management, MTB College Campus, Athwalines, Surat, Gujarat; phone: 9898718079; e-: nag.amrita@gmail.com

Nautiyal, Nanda. Department of Environmental Science, College of Basic Sciences and Humanities, G.B. Pant University of Agriculture and Technology, Pantnagar – 263145, Uttarakhand, India; e-: nautynan@gmail.com

Pant, SC. Govt. Degree College, Gairsain, Chamoli (H.N.B. University Garhwal, Srinagar) Uttarakhand

Patel, Pooja P. SR Luthra Institute of Management, MTB College Campus, Athwalines, Surat, Gujarat; phone: 9825964587; e-: pooja_patel1313@rediffmail.com

Qaisar, KN. Assoc. Prof. Faculty of forestry, Post Graduate in Forestry, Sher-E-Kashmir, University of Agricultural Sciences and Technology of Kashmir, Shalimar, Srinagar, India-191121

Rahim, MA. Professor, Department of Horticulture, Bangladesh Agricultural University, Mymensingh

Rastogi, Akanksha. Department of Environmental Science, College of Basic Sciences and Humanities, G.B. Pant University of Agriculture and Technology, Pantnagar-263145, Uttarakhand; e-: rastogi34002@gmail.com

Ravikumar, B.S. Department of Botany and Environmental Studies, AVK College for Women, Hassan, Karnataka- 573 201, India; e-: prakum2001@yahoo.co.in

Shankhdhar, Deepti. Department of Plant Physiology, College of Basic Sciences and Humanities, G.B. Pant University of Agriculture and Technology, Pantnagar-263145, Uttarakhand, India

Shankhdhar, S.C. Department of Plant Physiology, College of Basic Sciences and Humanities, G.B. Pant University of Agriculture and Technology, Pantnagar-263145, Uttarakhand, India; e-: shankhdhar.sc@rediffmail.com

Shiva, Vandana. Navdanya, 105, Rajpur Road, Dehradun – 248001, India

Shukla, Alok. Department of Plant Physiology, College of Basic Sciences and Humanities, G.B. Pant University of Agriculture and Technology Pantnagar, Uttarakhand – 263145, India

Singh, CP. KVK, Dhamora, S.V.P.U.A & T, Meerut, Rampur

Singh, Hukum. Department of Plant Physiology, College of Basic Sciences and Humanities, G.B. Pant University of Agriculture and Technology Pantnagar, Uttarakhand – 263145, India; e-: hukumsingh97@yahoo.com

Singh, Piyusha. Department of Genetics and Plant Breeding, College of Agriculture, G.B. Pant University of Agriculture and Technology, Pantnagar, U.S. Nagar, Uttarakhand- 263145; e-: piyusha_singh@yahoo.com

Singh, Prabjeet. College of Fisheries. GB Pant University of Agriculture and Technology, Pantnagar – 263145, Uttarakhand, India

Singh, Vir. Department of Environmental Science, College of Basic Sciences and Humanities, G.B. Pant University of Agriculture and Technology, Pantnagar – 263145, Uttarakhand, India; e-: drvirsingh@rediffmail.com, sinvir@gmail.com

Srivastava, MK. Head, Deptt of Plant Breeding and Genetics, ND College Shikohabad; e-: pbg_manoj@yahoo.co.in

Verma, Amit. Department of Biochemistry, G.B. Pant University of Agriculture and Technology, Pantnagar, U.S. Nagar, Uttarakhand- 263145; e-: amibiochem@rediffmail.com

Editorial

Biodiversity, Biosphere and Biological Bloom: Prerequisites for Cooling the Earth

☆ *Vir Singh, Nanda Nautiyal, Akanksha Rastogi and Neelendra K. Joshi*

Responsibility for life at large obliges us to support diversity, for diversity is not only the spice of life, it is also a precondition of vibrant and healthy life; moreover, a precondition of the meaningfulness and richness of our individual lives.

–Henryk Skolimowski

The earth's genes, species and ecosystems are the product of over 3000 million years of evolution, and are the basis for the survival of our own species.

–Jeffrey A. McNeely

Origin of life on Earth led to the creation of biosphere, a new feat of cosmic evolution. Life in the beginning had to cope with vulnerability due to only a few living forms. Therefore, evolution helped create more forms of life, higher forms of life, advanced forms of life, more evolved forms of life. The life continued evolving into a variety of species, into a variety of genotypes within species, into a variety of ecosystems. Thus, a diversity of life began to reign all over and the biosphere blossomed with innumerable ecosystems, species of plants, animals and microorganisms, and numerous varieties within species. The biosphere revolutionized with biodiversity and the planet Earth reverberated with fullness of life.

Biological Bloom: The Climax of Biodiversity

Life distinguishes Earth from other planets. Cosmic evolution has been especially benevolent to the planet Earth. Earth's biosphere has become a home for biodiversity. And biodiversity keeps the biosphere alive with life. The higher the degree of biodiversity the more the biosphere reverberates with life. The more the biosphere reverberates with life, the greater the chances of continuous biological bloom in the biosphere. Biological bloom means the greatest possible degree of biodiversity continually blossoming in the biosphere. The state of the biosphere being in biological bloom is a must for dynamic equilibrium and consequently for sustainability of life. The sustainable biosphere is continually nourished by the cosmic evolution by enriching the biodiversity by adding more and more species, more and more genotypes of species, and by creating more and more ecosystems and the complex ecological phenomena in the biosphere.

All is One: The Integrity of Life

Biological diversity, or as it is also called, biodiversity, encompasses all species of plants, animals, and microorganisms and the ecological processes of which they are parts. Or: the variety and variability among living organisms and the ecological complexes in which they occur. It is difficult, rather impossible, to enumerate the entire biodiversity on Earth. Importance of biodiversity, in fact, does not lie in mere number of species and the genetic information contained in them. The critical importance lies in the integrated totality of life, which is not there in the total count of species and their genotypes, and ecosystems. None of the ecosystems or species is independent or isolated from each other. All of them are one, constituting life, all ecosystems and species being dependent on each other and collectively contributing to the phenomenon making the planet Earth a living planet. Biodiversity makes a complex mosaic of structure and phenomena of life which cannot be seen or observed or researched. We can only get a feel of it by becoming an integral living element of it.

Biodiversity Count: Life is More than That

Globally, 1.4 million life forms have been named and described by taxonomists. However, as the biological estimates suggest, the total number of species might be 10 times those described so far. It might be even a tip of an iceberg. Life is more than biology can describe. Plant and animal species from major families are still being discovered in rainforests. Some new amphibians have recently been reported from Sathyamangalam forest in Tamil Nadu. Over 300 new fish species have come into picture and described from the Amazon region. About 15,000 to 20,000 species of tropical flowering plants are yet to be documented.

India is also a paradise of biodiversity. It is one of the six Vavilonian centres of origin of species and one of the world's 12 mega-biodiversity centres. Some 45,000 plant species and 89,000 animal species have been documented in India, which comprise some 6.5 percent of all known wildlife. Indian land is also credited with two biodiversity hot spots – the Eastern Himalaya and the Western Ghats – which unfortunately are amongst the world's most threatened hot spots. In fact, at least 10 percent of India's recorded wild flora and possibly more of its wild fauna are on the list of threatened species; many are on the brink of obliteration.

Human interest and craze for biodiversity seems confined to comprehend biodiversity by counting, hunting and utilising species and having records of endangered, near extinct and extinct species. More recent craze lies in genetic mapping of the species. But such a thoughtless dealing with biodiversity is detrimental and life-annihilating. Biodiversity is important only for a superficial idea about biodiversity. Life, in reality, is more than the totality of biodiversity.

All the requirements indispensable for the sustenance of life are fulfilled by the environment which is a biophysical entity as well as a phenomenon of biodiversity itself. The entire biodiversity, on the other hand, is a flowering of the environment as it is. Thus, biodiversity is owing to biospehere and vice-versa. The possibility of a biological bloom lies in the combined effect of environment and biodiversity, that is, in biosphere. Biosphere will be unimaginable without biodiversity. And biodiversity has no base to exist and be accommodated without biosphere in its intact and integrated form.

Biodiversity rules over all the components of environment, *viz.* lithosphere, hydrosphere and atmosphere. It is owing to living organisms – plants, animals and microorganisms – that parts of these environmental segments transform into biosphere.

Biodiversity and Ecological Integrity: A Prerequisite for Sustainability

Ecological integrity is the basic principle of sustainability of a production system such as agriculture, of the society as a whole, of the whole life, as well as of the living planet. It is the biodiversity that maintains this vital ecological integrity. This integrity is ensured by constant flows of matter contained in the environment (nutrients, liquids and gases) and energy. It is the biodiversity that maintains these flows. First, cosmic energy is trapped by green plants and then transformed into biochemical energy; it flows amongst organisms at various strata and also keeps transforming into other forms of energy. Had there been no biodiversity, such a transformation and flow of energy would have been impossible. It is the regulation of this flow that the environment stays in isothermal state (with little and comfortable range of fluctuations), significant reduction and rise in atmospheric temperature is avoided and factors and processes conducive to life are regulated.

Biodiversity helps maintain flow of nutrients in an ecosystem. Trees with the help of their deep root system draw valuable nutrients from the deeper layers of lithosphere. Again nutrients released after the decomposition of dead plant and animal bodies are recycled into an ecosystem. Nutrient flow also takes place through food chain. Green plants pick up nutrients from lithosphere, hydrosphere and atmosphere and also release these nutrients into these environmental components.

Biodiversity for Mitigating Climate Change

Biodiversity plays vital role in the recycling of water, which is also necessary for the distribution and flowering of life in all the components of the living environment. Water is not only necessary but a must for life, and recycling makes it available to all environments of the biosphere. Water (or hydrological) cycle not only nourishes and sustains life (water along with carbon dioxide transforms in glucose, the energy of life, by green plants in the presence of sunlight), but also regulates atmospheric temperature.

Biodiversity serves to utilize and contribute to atmospheric gases. Gaseous flow maintained in the biosphere through biodiversity is of vital repercussions for the living species. Microorganisms through nitrification utilize molecular nitrogen of the atmosphere. Plants utilize this nitrogen for protein synthesis. Animals consume plant nitrogen in the form of proteins. Microorganisms again release the nitrogen back into the atmospheric pool through what is called denitrification. Carbon dioxide is removed from the atmosphere through photosynthesis and added to the atmosphere through repiration. Plants produce oxygen by splitting water molecules. Thus, living organisms' role in the recycling of gases is well understood and is of phenomenal consequences in keeping the environment in a state of dynamic equilibrium.

Green plants are pivotal in removing carbon from the atmosphere through photosynthesis, a process called carbon sequestration. Carbon sequestration contributes to maintain carbon balance in the atmosphere which is vital for maintaining an appropriate temperature of the planet. Carbon dioxide occurs in the atmosphere only in traces (0.034 per cent) but it is responsible for the global warming to the extent of 62 per cent. Therefore, its increasing concentration in the atmosphere is of critical significance to the phenomenon of global warming and associated weather calamities, floods, untimely and uneven distribution of rains, storms, accelerated glacier melt, rise in sea level, high rate of species extinction, increasing incidence of diseases, etc.

Various ecosystems and their soils and vegetation types have their own efficiencies of carbon sequestration. Anthropogenic interference leading to deterioration of ecosystems by means of deforestatation, vegetation thinning, tree extraction, monocultures, etc. results in decreased efficiency of carbon sequestration and greater accumulation of carbon dioxide in the atmosphere towards the culmination of global warming and consequent climate change.

A forest ecosystem blossoming with biodiversity, including a variety of trees, shrubs, herbs, creepers, etc. and a variety of wild animals would naturally have its highest capacity to sequester carbon and contribute to mitigate climate change.

Wild animals add to a significant component of biodiversity. They contribute to enhance nutrient flows, seed dispersal, vegetation propagation, soil fertility, higher density and greater variety of vegetation, and thus contribute to the very essential ecological integrity of life. Biodiversity in wilderness in most of the cases is of highest order and phenomenally contributes to mitigate adverse climate change leading to cooling of the Earth.

Loss of Biodiversity = Warming of the Earth = Annihilation of Life

The Earth has been reverberating with fullness of life as it has been cool enough, on an average (average temperature of the planet being little less than 15°C), to support and sustain life. Even today the living planet continues to be cool on an average. It is not that the planet will cease to be cool in a few decades from now. The major concern is this that the phenomena responsible for keeping the coolness are gradually and gradually getting knocked down with adverse changes in the composition of biosphere which appears to gradually put the living planet into a state of insecurity and unsustainability. The coolness of the Earth is gradually losing with the loss of biodiversity, and it will not be wrong to say that the loss in coolness is proportional to the loss of biodiversity of the Earth.

In order to combat climate change characterized by global warming world's focus is on cutting down carbon emissions. Cutting down carbon emissions is one of the critical tactics for climate change mitigation. However, accelerating carbon sequestration is an equally important strategy to reverse climate change. Impact of carbon emissions becomes more pronounced if the carbon sinks are deteriorated and carbon sequestration process is hampered. On the other hands, if the carbon sinks are active, the impact of carbon emissions is minimized. The carbon sinks are in active state when the ecosystems – terrestrial as well as aquatic – flourish with biodiversity.

Several factors responsible for the global warming and climate change have been analysed by all sections of scientists and institutes.

However, loss of biodiversity as a major reason of the dismal state of the planet is often overlooked. Deterioration of the planet, of which global warming and climate change are the indicators, apparently is rooted in the accelerated loss of biodiversity. Loss to biodiversity to an irrepairable state would trigger the process leading to annihilation of life on Earth. Warming of the globe and annihilation of life would be proportional to the shrinkage of biodiversity.

Alternate Energy for the Amelioration of Life

Carbon emission can be prevented, not only controlled, with increasing dependence on alternative sources of energy. The world largely depends, virtually for almost all socioeconomic activities, on fossil fuels which are finite and pollution-creating energy resources. These conventional energy resources are a characteristic of the conventional development paradigm. The wealthiest nations are those who possess fossil fuels and have control over them. Development processes dependent on these energy resources which are responsible for critical level of carbon emissions leading to accumulation of free carbon in the atmosphere are not sustainable in real sense.

Sustainable development has two essential features: One, it is perfectly compatible with environment, and two, it does not compromise with the abilities of future generations. Fossil fuel energy dependent development approach pollutes environment, and future generations would be deprived of the finite energy resources in foreseeable future. Therefore, the conventional development is unsustainable. Therefore, a paradigm shift in the development is not only necessary but an imperative of our destiny.

All sources of energy on Earth emanate from the ultimate source of all energies, the solar energy. The solar energy, however, is not materialized to an appreciable extent. Biodiversity itself is the living product of solar energy. If solar energy could be used at a significant scale, it could do away all kind of pollution and cure other environmental and socioeconomic ills. There are other sources of energy derivable from inexhaustible resources, *e.g.* wind energy, geothermal energy, tidal energy, hydroenergy from run-of-the-river, etc. These energy sources would change the face of the world, help protect and conserve biodiversity, and be instrumental in ameliorating life on Earth.

The Biodiversity-Ecological Integrity-Sustainability Connections

Sustainable development is unimaginable without ecological integrity, which ranges from the ecological integrity of small production systems (e.g., agro-ecosystems, industrial areas, etc.) to that of the whole biosphere. It means, in order to operationalise, the sustainable development needs execution of three principles, *i.e.*, biodiversity, living soil and nutrient flows. Amongst these, the biodiversity serves as a key principle to create an eco-sphere and meet conditions for sustainability. The phenomenon of sustainability of life can be accommodated only in the ecosphere. Biodiversity creates living soil and ensures nutrient flows, which are indispensable for ecological integrity. Therefore, biodiversity – the basis of life on Earth – is the first and the ultimate factor to strike ecological integrity and sustainability.

Biodiversity is at the heart of sustainable development, as we have noted above. It takes care of the sustainability of all production systems as well as of the whole biosphere. In other words, biodiversity in its highest possible degree ensures the vital ecological integrity of the biosphere, and this is phenomenal for the very functional sustainability of the living planet, our Earth. A certain climate order vital for keeping life in the mode of enhancement is an outcome of this functional sustainability of the biosphere.

Biodiversity for Cosmic Balance

Crises of biodiversity are the crises of Earth. Should we need to reverse environmental crises, biodiversity is the key. Should we seriously attempt to mitigate climate change, biodiversity offers the opportunities. Should we be concerned about saving our living planet then biodiversity is the final mantra. Biodiversity makes biosphere. Biosphere accommodates biodiversity. Biodiversity gets nurtured and evolves in the biosphere. Biodiversity erosion at the hands of human species leaves scars on biosphere and then biosphere becomes less and less favourable for biodiversity. If biodiversity shrinks, biosphere would be at the verge of collapse.

Everything is related with everything else in this universe. Nothing exists, and nothing can exist, in isolation. Living planet strikes a balance in the cosmos by combating cosmic entropy. The living planet itself reverberates in perfect ecological integrity. The

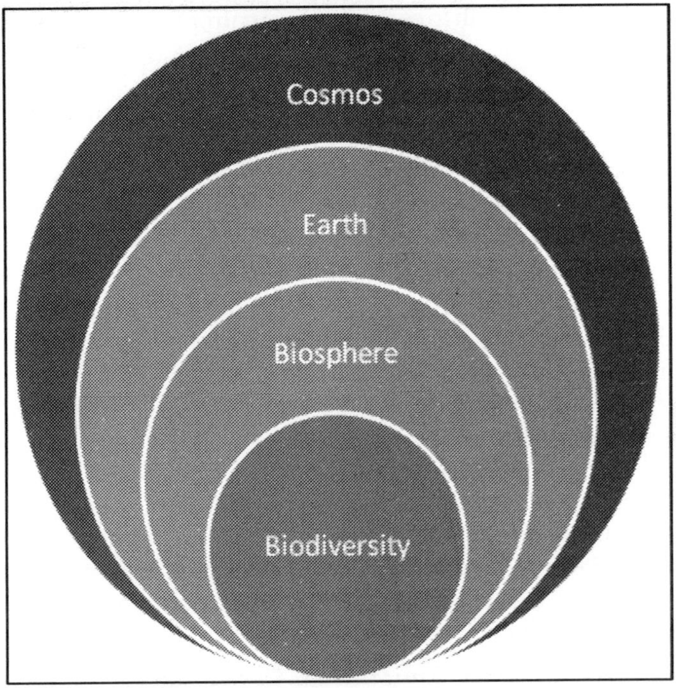

Figure 1.1: Biodiversity Blossoming in the Biosphere Ensures Ecological Integrity of the Living Planet. Living Earth combats cosmic entropy and phenomenally contributes to cosmic balance.

ecological integrity is a phenomenon of life which is owing to biodiversity imbibing on the living planet (Figure 1.1). Biodiversity thus is the supreme creation of the cosmic evolution. The Earth is a living planet owing to biodiversity. The Universe is a living universe owing to biodiversity.

Biodiversity in its fully blossoming state is a precondition for a healthy biosphere and vice-versa. Continually evolving and sustainably blossoming biodiversity and a healthy biosphere are precondition for a biological bloom on Earth. Biological bloom is to help the Earth combat cosmic entropy and thus contribute to the cosmic balance for which the living planet had come into existence. And only a cool Earth would be able to heal up the trauma caused by the ominous climate change. Cooling of the Earth would be inevitable for the Earth to combat cosmic entropy and strike a cosmic balance.

Conserving and Enhancing Biodiversity

Attempts to conserve biodiversity have been made the world over. Protected Areas are regarded to be the best bet for the conservation of biodiversity. Hill tribes have especially been conscious about biodiversity conservation. They demarcated certain natural areas imbibing with biodiversity as Sacred Groves. These Sacred groves have been serving as very safe and nourishing refuge for biodiversity for centuries. The traditional concept and approach of biodiversity conservation through Sacred Groves show the way how the biodiversity conservation should be translated into reality. The earliest protected areas with the help of public sector were established as wildlife sanctuaries in Asam in 1928 and 1934. In the northern Himalayan states of India, a network of protected areas was established in 1970s and 1980s, which included Jim Corbett National Park and Rajaji National park. In Nepal, 21 protected areas cover at least 26,666 km^2 of land. Chitwan was the first national park established in Nepal.

In Nepal, the Annapurna Conservation Area, the Kanchenjunga Conservation Area and the Makalu-Barun National Park are all run through community-based biodiversity management. In Bhutan the protected area system was established as early as 1960s. The Jigme Dorji Wangchuck National Park was the dominant component of the protected area system. In 1995, the protected area system was revised. The current protected areas in Bhutan, one of the countries having largest proportion of its geographical area under natural forest cover, has five national parks, three wildlife sanctuaries, and one strict nature reserve, accounting for about 26 percent of the total land area in Bhutan. In 1999, another 9 percent of the area in Bhutan was added to the protected system in the form of 12 biological corridors, which linked the protected areas to create a conservation landscapes extending across the country.

It is the nature that protects biodiversity. It is the biodiversity that makes nature. A healthy and vibrant nature embracing biodiversity at its full bloom will be an imperative for the Earth to restore its climate order and saving the living planet. Policy-based efforts of the governments for the conservation of biodiversity are important, but not adequate and cannot fully guarantee sustainability of such protected areas. People's participation in such protection strategies is almost negligible. What is most important is the eco-philosophy which inculcates amongst humans all the values necessary for the conservation and the enhancement of nature' biodiversity. Eco-philosophy created by Professor Henryk Skolimowski provides answers to all the problems pervading the planet Earth. It is a life-enhancing philosophy teaching reverential attitude for nature and providing ways and means of healing the Earth and healing the humanity. It attempts to empower every human person for the essential reconstruction of ecological world. It shows us the creative potential of our mind, which he calls the participatory mind.

Biodiversity and biosphere of the living planet are the creation of cosmic evolution. If we do not take evolution seriously we are doing disservice to the world and abusing our creativity-oriented mind. Prof. Henryk Skolimowski writes, "If we accept evolution continually transcending itself, in other words as emergent; and if we accept ourselves as a part of the evolutionary process, the question is, what values can we extract from the evolutionary process and how can these values, if judiciously applied, shape our attitude and behaviour?" Skolimowski suggests the following precepts extricated from intelligent reading of evolution:

- ☆ Behave in such a way as to preserve and enhance the unfolding of evolution and all its riches;
- ☆ Behave in such a way as to preserve and enhance life, which is a necessary condition for carrying on evolution;
- ☆ Behave in such a way as to preserve and enhance the ecosystem, which is a necessary condition for further enhancement of life and consciousness;
- ☆ Behave in such a way as to preserve and enhance the capacities which are the highest developed form of the evolved universe, consciousness, creativeness, compassion; and
- ☆ Behave in such a way as to preserve and enhance human life which is the vessel in which the most precious achievements of evolution are contained.

Biological bloom of the biosphere will be the climax of life on Earth. It is possible when the biodiversity blossoms at its fullness. Transforming our lifestyles in accordance with the conservation and enhancement of biodiversity will be the most appropriate tactic for us to keep our Earth to be in the state of biological bloom. Eco-philosophy offers to empower human beings to keep our living Earth in a state of biological bloom.

Part I

Climate Change: Risks and Vulnerability to the Biodiversity

Chapter 1

Biodiversity and Climate Change Mitigation

☆ *Mumtaz Majeed, Altamash Bashir, M.A. Khan and K.N. Qaisar*

Forests currently absorb billions of tons of CO_2 globally every year, an economic subsidy worth hundreds of billions of dollars if an equivalent sink had to be created in other ways. Concerns about the permanency of forest carbon stocks, difficulties in quantifying stock changes, and the threat of environmental and socioeconomic impacts of large-scale reforestation programs have limited the uptake of forestry activities in climate policies. With political will and the involvement of tropical regions, forests can contribute to climate change protection through carbon sequestration as well as offering economic, environmental, and sociocultural benefits. A key opportunity in tropical regions is the reduction of carbon emissions from deforestation and degradation. Forest community reacts and influences its environment. The intangible benefits from forests are due to these reactions. The influence of forest is not limited to the forest boundary, but some time it is observed far away as well. Forest has come to be defined as an area set aside or maintained under vegetation for any indirect benefits, namely climatic, protective, or environmental and/or for production of wood and non-wood products. The influence of forests on certain climatic parameters such as radiation, temperature, precipitation, humidity, wind movement etc. may lead to some change in macroclimate or create microclimate in the zone of its occurrence and surrounding areas. Global warming is the greatest crisis ever collectively faced by human kind, unlike other crisis; it is global in nature, threats the very survival of civilization and promises to throw up only losers over the entire international socio-economic fabric. The latest estimate by the Intergovernmental Panel on Climate Change (IPCC) shows that a business as usual scenario will lead to an increase in global mean temperature of about 1°C above the present value by the year 2025, and 3°C before the end of next century. IPCC estimates, about 300 million people would be severely affected, by a one meter rise in sea level due to the effect of deglaciation of Antarctic ice and consequent submergence of the low-lying coastal areas. Forests are more efficient in absorbing carbon dioxide than any other form of vegetation. Therefore these are usually called carbon dioxide sinks. Climate changes occur externally (from extra-terrestrial factors, solar out put, earth-sun geometry and stellar dust) or internally (from ocean, atmosphere and land-atmospheric chemistry, atmospheric albedo, surface albedo, ocean heat change, continental drift, mountain building and volcanic activity) through anyone of these components.

Introduction

Forests are a striking feature of land surface. The forests of a country are a natural asset of great value, which unlike mineral resources are renewable and can be perpetually protected, protective and useful under proper management. A forest community reacts and influences its environment. The influence of forests is not limited to forest boundary, but it is observed far away as well. The influence of forest on climate, soil, water, microclimate and biota are of great significance. The influences of forests on their environment are primarily due to:

1. The unbroken canopy consisting of fully stocked forest crop which intercepts sun-rays and precipitation, retards wind velocity and loss of heat by radiation and creates its own microclimate,
2. A large quantity of litter fills on the ground which protects and influences soil properties and hydrology to a great extent, and

3. Interlocked root system of trees which binds the soil together.

Forests of sufficient area and quality balance the quantity of carbon dioxide in the atmosphere. Continuous degradation of forest in different parts of the world has led to continuous increase in the earth temperature, disturbance in climate and several other adverse influences.

Global warming is the greatest crisis very collectively faced by the human-kind; unlike other crisis, it is global in nature, threatens the very survival of civilization and promises to throw up only losers over the entire international socio-economic fabric.

There are many greenhouse gases which could adversely affect the world's climate but most steadily increasing gases are carbon dioxide, methane, chlorofluorocarbons and nitrogen oxides.

Forest and Climate

The macro-climate of a given area or region is determined primarily by geographical location, latitudes and altitudes. The influence of forests on certain climatic parameters, such as radiation, temperature, precipitation, humidity, wind movement etc, may lead to some change in the macro-climate or create micro-climate in the zone of its occurrence and surrounding areas (Figure 1.1).

Forest and Global Carbon Dioxide Balance and Global Warming

The carbon dioxide occupies about 350ppm of the atmosphere. It is believed that concentration of carbon dioxide during the industrial era, *i.e.* before 20th century, was not more than 295ppm. Seasonal variation of about 6ppm in the concentration of carbon dioxide is caused due to consumption of carbon dioxide during the growth period of plants and due to respiration of vegetation. The seasonal and daily variation is more pronounced just above the vegetation layer in the atmosphere.

A forest plays an important role in fixing carbon dioxide through the process of photosynthesis and is represented by the equation as:

$$6CO_2+6H_2O+675\text{Kcalories} \rightarrow C_6H_{12}O_6+6O_2$$

In order to produce 1gm of dry matter, 1.69 g of carbon dioxide assimilation is required. The process involves simultaneous release

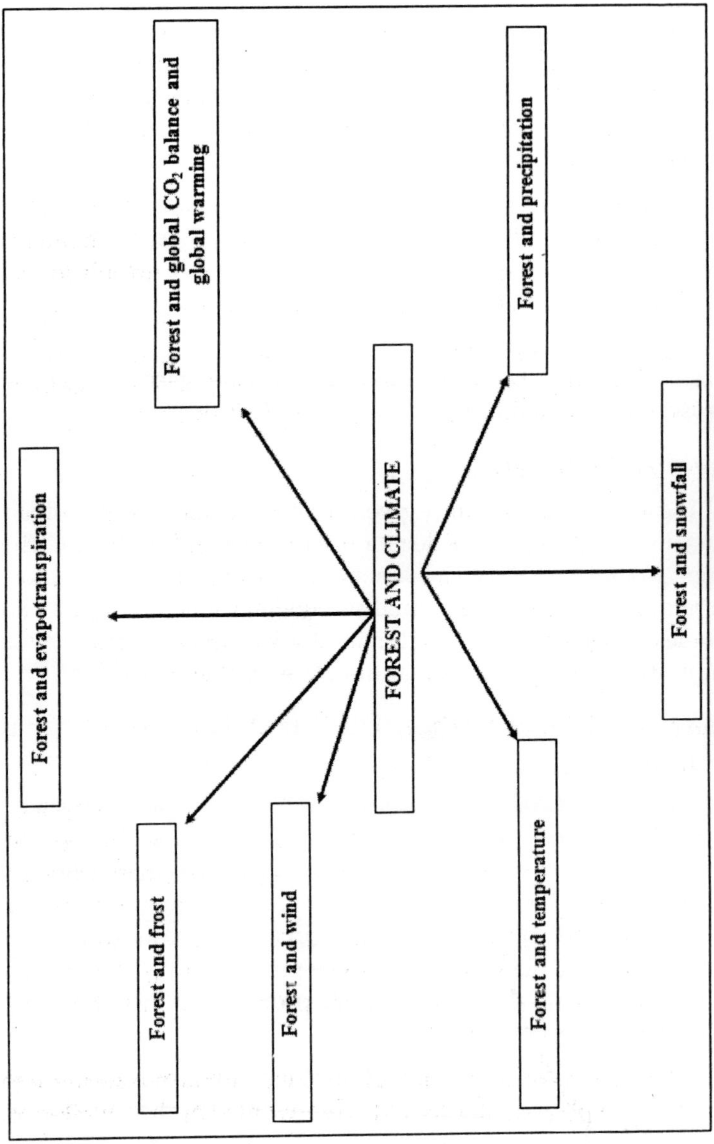

Figure 1.1: Forests and Climate

of 1.22 g of oxygen. Forests are more efficient in absorbing carbon dioxide than any other form of vegetation. Therefore, these are usually called carbon dioxide sinks.

Forest and Precipitation

It is believed that forests increase precipitation and destruction of forest leads to reduction on precipitation. Repeated occurrence of drought conditions in the country are attributed to the loss of forests.

Forest and Temperature

Forests exercise a moderating influence on air temperature. Although there is a considerable variation in the data obtained regarding effect of forests on temperature, the evidence is conclusive that forests reduce air temperature.

Forest and Wind

Forest offers mechanical obstruction and deflects upwards a large part of moving mass of air and slows down the velocity of wind which enters the forests.

Forest and Frost

Forests have moderating influences on air and ground temperature. Forests work as a blanket and do not allow temperature to fall considerably to cause frost.

Forest and Snowfall

Forests affect snowfall and snow melting. Forested area contains almost twice snow-water than deforested areas and it delays melting of snow.

Forest and Evapo-transpiration

Due to moderating influence of forest on solar radiation, temperature and wind, forest reduces evaporation of moisture from surface soil. The effect may vary with type, age, leaf-litter, density of forest and moisture in the soil. Evaporation of rainfall from forest floor may be as low as 10 per cent and is high as 80 per cent of that from bare soil (Champion and Seth, 1968).

It is certainly logical to assume that temperature in the lower levels of the forest differ more from external temperatures than those taken at the level of the canopy of the dominant species. Atmospheric humidity should vary according to the same law; and similarly, light should lose some of its intensity – while yet remaining very bright – according to the varying density and structure of the levels of vegetation. One of the few comparative observations was carried out by Carton and Sallenave in the experimental forest of Trang-Bom in South Vietnam. Some of the results of the observation are shown in Tables 1.1–1.3.

Table 1.1: Air Temperature

	In Open	*Under Forest*	*Difference*
		Centigrade	
Annual mean of maxima	32.1	**28.4**	3.7
Annual mean of minima	21.4	**22.5**	1.1
Difference	10.7	**5.9**	4.8
Maximum diurnal range	18.5	**13.5**	5.0
Minimum	1.0	**0.03**	0.7
Difference between extreme maxima and extreme minima for whole period	27.7	**24.0**	3.7

Table 1.2: Relative Humidity

	In Open	*Under Forest*	*Difference*
		Percentage	
Mean annual at 6 a.m.	97.9	96.7	1.2
Mean annual at 2 p.m	67.1	82.3	15.2
Difference	30.8	14.4	16.4
Greatest difference between mean monthly at 6 a.m and 2 p.m	51.8	42.9	8.9
Smallest difference between mean monthly at 6 a.m and 2 p.m	10.7	2.7	8.0
Difference between extreme maxima and extreme minima at 2 p.m for whole of period recorded	**77.0**	**72.0**	**5.0**

Table 1.3: Relative Humidity

	In Open	*Under Forest*	*Difference*
		Millimetres	
Total for mean year	**702.2**	**297.5**	**404.7**
Highest monthly total	126.4	67.1	-59.3
Lowest monthly total	26.2	6.6	-19.6
Mean total per 24 hours	1.9	0.8	1.1
Maximum per 24 hours in whole period	8.3		

Forest and Interception

Interception refers to that part of precipitation which is caught by vegetative cover, *e.g.* tree crowns, under storey vegetation, etc. The intercepted precipitation is either evaporated or distributed to the soil through stem flow (Table 1.4).

Table 1.4: Through Fall, Stem Flow and Interception Under Different Forest Covers

Species	*Stand Density Stems/ha*	*Through Fall (mm)*	*Stem Flow (mm)*	*Interception (mm)*
Shorea robusta	668	54.6	7.2	**38.2**
Shorea robusta	1664	**81.0**	6.9	13.1
Shorea robusta	1678	66.4	8.3	25.3
Shorea robusta	2648	72.4	9.6	18.0
Alstonia scholaris	1675	57.0	**17.0**	26.0
Pinus roxburghii	1156	74.3	3.6	22.1
Tectona grandis	475	73.2	6.0	20.8
Acacia catechu	574	73.2	6.0	20.8
Eucalyptus hybrid	1658	80.8	7.7	11.5

Forest and Erosion

Forest protects soil from water and wind erosion. The soil loss from areas under natural forest cover is very small or almost negligible as compared to the soil loss under cultivated, fallow, dry land agriculture, agri-horticulture and grass cover (Anonymous, 1976).

Soil erosion measurements carried out under different conditions in U.S.A. indicated minimum erosion under forest cover (Table 1.5).

Table 1.5: Indicated Time Necessary to Erode 17.5 cm of Soil Material from Cocil Clay Loam at 10 per cent Slope in USA

Treatment	*Annual Soil Loss (t/ha)*	*Years Required to Erode 17.5 cm Soil*
Virgin forest	**0.005**	**500000**
Grass	0.77	3225
Rotation crops	35.70	70
Cotton	78.05	32
Bare ground	**165.50**	**15**

The Atmosphere

The atmosphere is a mixture of different gases and aerosols (suspended liquid and soil particles) collectively known as air which envelops the Earth, forming an integrated environmental (climate) system with all the Earth's components. The atmosphere provides various functions, not least the ability to sustain life. Of prime interest for a discussion on climate change, however, is its ability to control the Earth's energy budget. To understand this process, it will be necessary to study in more detail the composition of the atmosphere.

Composition

Let us consider first the atmospheric gases. Table 1.6 illustrates the average gaseous composition of dry air below 25km. Although traces of atmospheric gases have been detected well out into space, 99 per cent of the mass of the atmosphere lies below about 25-30km altitude, whilst 50 per cent is concentrated in the lowest 5km (less than the height of Mount Everest).

This gaseous mixture remains remarkably uniform in composition, and is the result of efficient biogeochemical recycling processes and turbulent mixing in the atmosphere. The two most abundant gases are nitrogen (78 per cent by volume) and oxygen (21 per cent), and together they make up over 99 per cent of the lower atmosphere.

Table 1.6: Average Composition of the Atmosphere below 25 km

Component	*Chemical Abbreviation*	*Volume % (Dry Air)*
Nitrogen	N_2	78.08
Oxygen	O_2	20.98
Argon	Ar	0.93
Carbon dioxide	CO_2	**0.035**
Neon	Ne	0.0018
Helium	He	0.005
Hydrogen	H	0.0006
Krypton	Kr	0.0011
Xenon	Xe	0.00009
Methane	CH_4	0.0017
Ozone	O_3	0.00006

Table 1.7: Contribution to Global Warming by Greenhouse Gases and Human Activities (% of global warming)

Sector	*Carbon Dioxide*	*Methane*	*Ozone*	*Nitrous Oxide*	*Chiorofluo-carbons*	*% Warming by Sector*
Energy						
Direct	35	3	X	**4**	X	**49**
Indirect	X	1	**6**	X	X	
Deforestation	10	4	X	X	X	**14**
Agriculture	3	**8**	X	2	X	13
Industry	2	X	2	X	**20**	24
% warming by gas	**50**	16	8	6	20	100

There is no evidence that the relative levels of these two gases are changing significantly over time. Despite their relative scarcity, the so-called greenhouse gases play an important role in the regulation of the atmosphere's energy budget.

Greenhouse Effect and Global Warming

There is an increasing concern about the gradual warming of the atmosphere through "greenhouse effect" and consequent climatic changes. The latest estimate by the Intergovernmental Panel on

Climate Change (IPCC) shows that a business as usual scenario will lead to an increase in global mean temperature of about 1°C above the present value by the year 2025 and 3°C before the end of the next century. There are many greenhouse gases (Table 1.7) which could adversely affect the world's climate but most steadily increasing gases are carbon dioxide, methane, chlorofluorocarbons and nitrogen oxides.

The atmosphere acts as a greenhouse for the Earth, letting through incoming light but shutting in heat. An increased amount of certain gases (carbon dioxide, water-vapours, methane, nitrous oxide, chlorofluoro carbons) increases the atmosphere's ability to block heat, which leads to warmer temperatures and climate changes. The greenhouse gases have peculiar optical properties. They are almost completely transparent to incoming solar radiation which in turn is remitted at longer wavelengths in the Infrared range by the earth's surface. Greenhouse gases absorb a good part of this long-wave radiation, thus warming the lower atmosphere of the earth (Mintzer, 1988).

An important event in the field of atmospheric chemistry was the realization that carbon dioxide was not the only infra-red absorbing trace constituent which was accumulating in the global atmosphere, other substances now recognized as greenhouse gases are methane, chlorofluorocarbons, nitrous oxide, ozone, carbon monoxide, hydrocarbons and sulphur compounds (Table 1.8). The preindustrial and 1989 concentration of some of these gases, with their increasing trend and possible sources of increases have been shown in Table 1.11. The changes from pre industrial times have been significant, with carbon dioxide increasing by 25 per cent, methane by more than 100 per cent and CFC's being totally man-made, not being present in the atmosphere prior to the 1930's. These observations show that the actual concentrations of different greenhouse gases vary by up to a factor 10 when comparing carbon dioxide and chlorofluorocarbons.

It has been estimated that carbon dioxide accounts for about 50 per cent of the global warming effect, chlorofluorocarbons 20 per cent, methane 12 per cent, nitrous oxide 7 per cent and other trace gases the remainder. During this era, carbon dioxide has contributed the most to global warming. In the future, the onus for global warming will rest more heavily on gases other than carbon dioxide.

Table 1.8: Greenhouse Gases, their Pre-industrial and Current Concentration with Possible Sources of Increase

Greenhouse Gas	*Concentration*		*Present Trend (% per y)*	*Possible Sources of Increases*
	Pre-indust (1850)	*Current (1989)*		
CO_2	275 ppmv	350 ppmv	0.4	Fossil fuel combustion, deforestion
CH_4	**750 ppbv**	**1750 ppbv**	**0.8**	Rice paddies, ruminants, biomass burning, gas and coal fields, land fills, tundra
CFC-11	Nil	250 pptv	4	Industrial and consumer goods
CFC-12	Nil	450 pptv	4	-do-
N_2O	**285 ppbv**	**310 ppbv**	**0.3**	**Biomass burning, agriculture, fossil fuel combustion**
O_3 (trop.)	15-20 ppbv	20-30 ppbv	0.5a	Urban and industrial pollution

The maps prepared by NRSA for the period 1972-75 and 1980-82 showed that in the seven years interval the extent of forest cover in the country had gone down by about 9.18 million hectares. The annual loss of forest cover during this period worked out to 1.3 million hectares. Similarly the FSI figures of 1981-83 and 1985-87 indicated reduction in forest cover of 0.19 million hectares during the intervening period of four years. The annual loss of forest cover worked out to 47,500 hectares, which is about one-thirtieth of what assessed for the previous decade by NRSA estimate, the area under closed forests had decreased and the area under open forests had increased, whereas exactly the opposite scenario is depicted by the FSI estimates. The details are given in Table 1.9.The results indicate that the annual deforestation rate in the country in the 1980's is much lower in comparison to what it was during the 1970's. However, in order to evaluate objectively the results of these studies it is important to understand the mechanism of changes in the forest cover on the ground and the potential of satellite remote sensing to detect and measure these changes.

Table 1.9: Changes in Forest Cover in Million Hectares

	Year/Category	Closed Forest	Open Forest	Mangrove	Total Forest Area
A.	NRSA estimate* (1972-75 to 1980-82)	(-)10-40	(+)0.07	**(+)0.07**	**(-)19.8**
B.	FSI estimate $ (1981-83 to 1985-87)	(+) 1.70	(-) 1.92	**(+) 0.03**	**(-) 0.19**

* Time interval: 7 yrs; Annual deforestation rate = 1.3 million ha.

$ Time interval: 4 years; Annual deforestation rate = 47,500 ha.

+: Increase in forest cover.

–: Decrease in forest cover.

Changes occur in forests due to various man-induced as well as natural causes such as removal of trees, shifting cultivation, fire, grazing, disease, insects, defoliation, and natural calamities. The term deforestation connotes forest depletion and forest degradation combined.

Consequences of Deforestation

Both selective and extensive deforestation brings about loss of biodiversity, depletion and extinction of valuable genetic resources, desertification of land, soil erosion, climatic change. The ecological inter-relationships are complex. Even a simplistic attempt to model some interrelationships begins to show that one cannot view environmental events in isolation.

Biodiversity

Countless species of plants and animals found in forest ecosystems become endangered or extinct with the destruction of their habitat. Biodiversity is the wealth of life on earth. The millions of plants, animals and microorganisms, the genes they contain and the intricate ecosystems they help build into the living environment are all included under the benevolent umbrella of biodiversity. Scientists have identified and classified some 1.4 million species, but millions more remain unknown.

Twenty five per cent of the medicines prescribed in the United States come from Tropical Rain Forest plants. Currently, about 120 prescription drugs that are used worldwide contain plant-derived ingredients from the rain forests. It is estimated that only 1 per cent of tropical plants have been tested for medical uses. This should be strong incentive to protect what remains of this potential medical resource.

Desertification

What is the first step along the road to desertification? In dry areas where vegetation is sparse, trees and open woodlands play a vital role in stabilizing soil and water. When the trees are removed, the land becomes more exposed to the elements.

Prediction of carbon dioxide emission from fossil fuel use in future is exonerated with uncertainty. It is estimated that global carbon dioxide emission from fossil fuel combustion in the year 2050 will be 2-20 Gt. Another estimate uses 1975 as the historical datum and projects a rate of increase of 0.07 Gt/year, giving carbon dioxide emissions in the year 2005 a value of 6.2 Gt (Mintzer, 1987). A third estimate may be obtained by assuming that the post 1973 average of growth of 1.46 per cent per year would continue in the future. The resultant figure for the year 2004-2005 is 7.0 Gt. In the year 2004-2005, carbon dioxide emissions from fossil fuels used in India, as a percentage of global emissions, would be 5.5-6.3 per cent. Table 1.10 summarizes the estimated fossil fuel and carbon dioxide emissions for India by the year 2004-2005.

Table 1.10: Estimated Carbon Dioxide Emission for India by 2004–2005

Fuel	*Predicted Fuel Requirement (Mt)*	*Carbon Dioxide Emission (Mt)*
Coal	576	**303.8**
Petroleum	92.44-101.13	72.1-78.88
Natural gas	**25**	**14**
Total		389.9-396.71

The Combined Effect

Calculating the combined effect of the various greenhouse gases is a fairly complex process. The impact of each gas depends not just on its greenhouse properties but on its lifetime in the atmosphere. The greenhouse effect of a certain amount of methane, for example, which is added to the atmosphere today, will decline relatively rapidly as the gas is broken down by natural processes. But similar amount of CFC's-12 will still be active in the early part of the twenty second century. The overall impact of a greenhouse gas, taking into account its radioactive effect and its lifetime, is referred to as its global warming potential.

The present concentration and rates of increase of the principal greenhouse gases in the atmosphere have been shown in Table 1.11. The table also shows the lifetime in the atmosphere and the global warming potential of each gas in comparison with carbon dioxide. The comparison is based upon the effect of 1 kg of each gas over a period of 20yrs.

Table 1.11: Comparison of Greenhouse Gases

Greenhouse Gas	*Approx. Greenhouse Impact Compared with Carbon Dioxide*	*Present Conc.*	*Present Rate of Increase (9% yr)*	*Lifetime in Atmosphere (yrs)*
Carbon dioxide	1	353 ppm	0.5	**50-200**
Methane	60	1.7 ppm	**0.9**	10
Nitrous oxide	270	0.3 ppm	**0.2**	150
Ozone	2000	100 ppb	1	**0.1**
CFC-11	4500	0.3 ppb	4	65
CFC-12	**7000**	0.5 ppb	4	130
HCFC-22	4100	NA	NA	15

The combined effect of the greenhouse gases so far added to the atmosphere as a result of human activities is equivalent to a 50 per cent rise in the carbon dioxide concentration. Of that about half is attributed to carbon dioxide itself and the remainder to the other greenhouse gases (Foley, 1991).

Conclusion

- ☆ Forests are the natural purifiers of air. They act as absorbers of greenhouse gases like CO_2, nitrous oxide, carbon particles etc. and releasers of O_2. They absorb poisonous gases like SO_2, NO_2, flourides, trace metals etc.
- ☆ Due to large-scale deforestation and rapid industrialisation and urbanisation, the world is heading

towards a major catastrophe in terms of global warming and climate change.

- ☆ These two phenomena, if not checked immediately, will affect 300 million people as 1.0m rise in sea level due to de-glaciation of Antarctic ice and subsequent submergence of low lying coastal area, according to IPCC.
- ☆ Therefore, reduction in the emission of greenhouse gases *viz.* CO_2, methane, nitrous oxide, CFCS etc. to a significant extent and also reforestation and afforestation programmes should be taken on war footing basis on global level.
- ☆ Hear the industrially developed nations of the world who have to play an active role in initiating the steps to reduce the emission of greenhouse gases rather than to blame each other and bear the consequences.
- ☆ In this respect, the Kyoto Protocol is a welcome step to which almost all the countries except US, Australia and some African countries are signatories and have ratified it. But US must take a sincere step to endorse and ratify Kyoto Protocol which was adopted in 1997, ratified and signed by most of developed and developing countries, because global warming is not region- or country-specific problem rather its consequences (which are detrimental) will be global and US will be no exception to that. The IPCC report that predicted warming is based on review of data by nearly 2,500 international scientists and climatologists.

References

Anonymous. 1976. *Soil and Water Conservation Research, 1956-71.* ICAR, New Delhi.

Champion, H.G. and Seth, S.K. 1968. *General Silviculture for India.* Government of India Publication, New Delhi.

Foley, G. 1991. *Global Warming: Who is Taking the Heat?* Panos Publications, Ltd, London.

Mintzer, I.M. 1987. *A Matter of Degrees: The Potential for Controlling the Greenhouse Effect.* World Resource Institute, Washington D.C.

Mintzer, I.M. 1988. A matter of degrees: A fresh look at the green house effect. *Environmental Awareness,* 11:3 and 4:49-55.

Chapter 2

Biodiversity of Submontane Forest in Tarai and Bhabar of Kumaun Himalaya

☆ *Bhasker Joshi and S.C. Pant*

Biodiversity of vegetation was analyzed in a submontane forest of Tarai and Bhabar of Kumaun adjacent to Kashipur, at (29° 14-43.6)–(29° 19-50.5) E longitude and (79° 03-22.6)–(79° 04-23.2) N latitude at an elevation of 253.4–265.5 meter asl. Forests provide oxygen by photosynthesis and consume huge amount of carbon dioxide. They reduce the effects of global warming. As many as 206 genera of dicotyledons, 44 genera of monocotyledons and 11 genera of pteridophytes were recorded in which 270 species of dicotyledons, 72 species of monocotyledons and 14 species of pteridophytes were identified. As many as 356 herbaceous species of 261 genera and 80 families were recorded and identified during study period. Construction of roads, urban expansion, settled agriculture and pastoralism have caused massive forest destruction in Tarai and Bhabar areas of Kumaun Himalaya. The forests also produce row materials for paper industries. Large numbers of paper industries are established in Tarai and Bhabar area. Due to anthropogenic factors the area under forests is shrinking rapidly. The effect of anthropogenic agencies in term of structure and function of forest ecosystem in Tarai and Bhabhar is still unknown. Thus detailed ecological study was carried out in the forest sites to understand the effect of biotic/anthropogenic agencies in t1he structure and function of forest ecosystem.

Introduction

The Himalayan forests have high degree of biodiversity and forest vegetation ranges from tropical dry deciduous forests in the foothills to alpine meadows above timberline (Singh and Singh, 1992). Champion and Seth (1965) and Beg (1975) carried out an observational survey and recognized various types of forests and different vegetational zones on the basis of temperature and altitude. Singh and Singh (1987) conducted the study of forest vegetation of the Himalaya with emphasis on: paleoecological, phytogeographical, and phytosociological aspects of vegetation; structural and functional features of forest ecosystems; and relationship between man and forests. Agrawal (1990) reported the analysis of floristic composition in Western Himalaya. Adhikari *et al.* (1992) and Saxena *et al.* (1985) reported that vegetation types differ with change in altitude. Kumari and Biswas (2003) studied forest biodiversity conservation in Indian Himalaya.

Species composition of major forest types of Central Himalaya is described by Tewari and Singh (1981), Ralhan *et al.* (1982), Singh *et al.* (1984), Singh and Singh (1984), Tiwari and Singh (1985), Upreti *et al.* (1985), Singh and Singh (1987), Singh *et al.* (1987), Rikhari *et al.* (1989), Adhikari *et al.* (1989), Bankoti (1990), Singh and Singh (1992) and Dhar *et al.* (1997). Structural and functional characteristics of individual forest ecosystem in Himalayan forests were conducted by Rawal *et al.* (1994), Rikhari *et al.* (1997), Kala and Uniyal (1999), Samant (1999), Dhar (2000), and Silori (2001). There is little information available for vegetational analysis in Tarai and Bhabar area of Kumaun by Pant (1976) and Jain and Sastry (1983). Pant *et al.* (1981) recorded 488 plant species in Corbett National Park.

The increasing population trend over the last few decades and high dependence of human beings on plant products have led to the vast exploitation of natural flora and fauna of this region. Changes in the land use, habitat fragmentation, nutrient enrichment and environmental stress often lead to reduced plant biodiversity in natural ecosystem (Mooney *et al.*, 1996, Chapin *et al.*, 1997 and Vitousek *et al.*, 1997). The increasing human activities on biodiversity might impact ecosystem functioning. Present study provides complete information on life forms, biological spectrum and vegetational analysis of forest vegetation in a part of Tarai and Bhabar of Kumaun Himalaya.

Geographical Location

For the present study, the forests of Tarai and Bhabar area of Kumaun Himalaya adjacent to Kashipur were selected. The selected forest area was 3242.09 ha (Source: Office of Tarai West Forest Division, Kumaun, Ramnagar, Uttarakhand). The study was conducted from April 2007 to March 2008. The study sites were situated in the foothills of Shivalik mountain of the Outer Himalaya and south-east to Corbett National Park at (29° 14-43.6)–(29° 19-50.5) E longitude and (79° 03-22.6)–(79° 04-23.2) N latitude at an elevation of 253.4–265.5 meter above the mean sea level, within the districts of Nainital and Udham Singh Nagar and occupied the middle reaches of the river Kosi and Dabaka.

Materials and Methods for Vegetational Analysis and Biological Spectrum

The plants were identified with the help of plant taxonomists and the forest flora of Kumaun (Osmoston, 1926), Flora Simlensis (Collet, 1971), Flora Nainitalensis (Gupta, 1968) and Flora of Mussoorie (Raizada, 1978).

Biological spectrum of study site was analyzed by methods described by Raunkiaer (1934) as phanerophyte, chamaephyte, hemicryptophyte, geophyte and therophyte.

The percentage life-form was calculated by following formula.

$$\%\text{ Life - Form} = \frac{\text{Number of species in any life form classes}}{\text{Total number of species of all life - form classes}} \times 100$$

Results and Discussion

Ramnagar range forest has high floristic diversity. In present study 261 genera and 356 plants species of 80 families were identified (Table 2.3) in which 234 herbs, 39 shrubs, 60 trees, 14 pteridophytes and 9 climbers were identified. Herbaceous species were classified as grasses, sedges, legumes and forbs. As many as 38 grasses, 11 sedges, 21 legumes and 164 forbs were identified during study period. In the study area 206 genera of dicotyledons, 44 genera of monocotyledons and 11 genera of pteridophytes were recorded, in which 270 species of dicotyledons, 72 of monocotyledons and 14 of pteridophytes were identified.

Table 2.1

Group	*Genera*	*Species*
Dicotyledons	206 or (78.92 %)	270 or (75.84 %)
Monocotyledons	44 or (16.85 %)	72 or (20.22 %)
Pteridophytes	11 or (4.21)	14 or (3.93 %)
Total	261 or (99.98 %)	356 or (99.98 %)

Table 2.3 shows life forms of plants in over all forest area. Table 2.4 summarises the categories of life forms. It consists of various life form classes like phanerophytes, chamaephytes, hemicryptophytes, cryptophytes and therophytes. The various life forms were recorded as 24.43 per cent, 3.08 per cent, 5.89 per cent, 6.74 per cent and 59.83 per cent respectively.

Table 2.2

	Ph	*Ch*	*He*	*G*	*Th*
Biological Spectrum of Study Site	24.43%	3.08%	5.89%	6.74%	59.83%
Total Plant Species	87	11	21	24	213
Raunkiaer's Normal Spectrum	46.00%	9.00%	26.00%	4.00%	13.00%

Ph: Phanerophyte; Ch: Chamaephyte; He: Hemicryptophyte; G: Geophyte; Th: Therophyte.

In the present study, maximum 291 species were recorded in Gulzarpur forest. Jurkha forest had 290 species, Aampani 236

Table 2.3: Life Form, Biological Spectrum and Occurrence of Species in each Forest in a Year

S.No.	Name of Plant Species	Family	Habit	Life Form	Herb Category	Ju*	Gu*	Jo*	Aa*
1.	*Abelmoschus moschatus* Medik.	Malvaceae	Herb	Th	Forb	+	–	–	+
2.	*Abutilon indicum* Sweet	Malvaceae	Herb	Th	Forb	+	+	+	+
3.	*Abutilon persicum* (Burm. f.) Merr	Malvaceae	Herb	Th	Forb	+	+	+	+
4.	*Acacia auriculaeformis* Cunn. ex Benth.	Mimosaceae	Tree	Ph		–	+	–	–
5.	*Acacia catechu* Willd.	Mimosaceae	Tree	Ph		+	+	+	+
6.	*Acacia farnesiana* Willd.	Mimosaceae	Shrub	Ph		+	+	+	+
7.	*Acacia nilotica* (L.) Willd.	Mimosaceae	Shrub	Ph		+	+	+	+
8.	*Achyranthes aspera* L.	Amaranthaceae	Herb	Ch	Forb	+	+	+	+
9.	*Adenostemma lavenia* L.	Asteraceae	Herb	Th	Forb	+	+	+	+
10.	*Adiantum capillus-veneris* L.	Adiantaceae	Fern	G		+	+	–	+
11.	*Adiantum incisum* Forssk.	Adiantaceae	Fern	G		+	+	–	+
12.	*Adiantum philippense* L.	Adiantaceae	Fern	G		–	–	+	+
13.	*Adina cordifolia* (Roxb.) Benth & Hook, f.	Rubiaceae	Tree	Ph		+	+	–	–
14.	*Aegle marmelos* (L.) Correa	Rutaceae	Tree	Ph		+	+	+	+
15.	*Aerva scandens* Wall.	Amaranthaceae	Herb	Ch	Forb	+	+	+	+
16.	*Aeschynomene indica* L.	Fabaceae	Herb	Th	Legume	–	–	+	–

Contd...

Table 2.3–Contd...

S.No.	Name of Plant Species	Family	Habit	Life Form	Herb Category	Ju*	Gu*	Jo*	Aa*
17.	*Ageratum conyzoides* L.	Asteraceae	Herb	Th	Forb	+	+	+	+
18.	*Ageratum haustonianum* Mill.	Asteraceae	Herb	Th	Forb	–	–	+	+
19.	*Ajuga bracteosa* Wall. ex Benth.	Lamiaceae	Herb	Th	Forb	+	+	–	–
20.	*Albizzia chinensis* (Oseck) Merr.	Mimosaceae	Tree	Ph		+	+	–	–
21.	*Albizzia procera* Benth.	Mimosaceae	Tree	Ph		+	+	+	+
22.	*Alternanthera sessilis* R.Br.	Amaranthaceae	Herb	Th	Forb	+	+	+	+
23.	*Alysicarpus hamosus* Edgew.	Fabaceae	Herb	Th	Legume	+	+	–	+
24.	*Alysicarpus vaginalis* DC.	Fabaceae	Herb	Th	Legume	+	+	+	+
25.	*Amaranthus virdis* L.	Amaranthaceae	Herb	Th	Forb	+	+	+	+
26.	*Anagallis arvensis* L.	Primulaceae	Herb	Th	Forb	+	+	+	+
27.	*Anisochilus carnosus* Wall.	Lamiaceae	Herb	Th	Forb	+	+	–	+
28.	*Anisomeles indica* L.	Lamiaceae	Herb	Th	Forb	+	+	+	+
29.	*Antirrhinum orontium* L.	Scrophulariaceae	Herb	Th	Forb	+	–	–	–
30.	*Argemone mexicana* L.	Papaveraceae	Herb	Th	Forb	+	+	+	+
31.	*Artemisia parviflora* Roxb. ex D.Don	Asteraceae	Herb	Th	Forb	–	+	–	–
32.	*Arthraxon lancifolius* (Trin.) Hochst	Poaceae	Herb	Th	Grass	+	+	+	+
33.	*Asperagus adscendens* Roxb.	Liliaceae	Herb	Th	Forb	–	+	–	–

Contd...

Table 2.3–Contd...

S.No.	Name of Plant Species	Family	Habit	Life Form	Herb Category	Ju*	Gu*	Jo*	Aa*
34.	*Asperagus racemosus* Willd.	Liliaceae	Herb	Th	Forb	–	–	+	–
35.	*Bambusa arundinaceae* Willd.	Poaceae	Tree	Ph		+	+	–	+
36.	*Barleria cristata* L.	Acanthaceae	Herb	Th	Forb	–	+	–	+
37.	*Bauhinia malabarica* Roxb.	Caesalpiniaceae	Tree	Ph		+	+	+	+
38.	*Bauhinia vahlii* W. & A.	Caesalpiniaceae	Climber	Ph		+	+	–	+
39.	*Bidens biternatea* Merr. & Sherf	Asteraceae	Herb	Th	Forb	+	+	+	+
40.	*Biophytum sensitivum* Zucc.	Oxalidaceae	Herb	Th	Forb	+	+	+	+
41.	*Blumea mollis* D.Don	Asteraceae	Herb	Th	Forb	+	+	+	+
42.	*Blumea oxyodonta* DC.	Asteraceae	Herb	Th	Forb	+	+	–	+
43.	*Boehmaria scabra* Gaud.	Urticaceae	Shrub	Ph		+	–	–	–
44.	*Boerhaavia diffusa* L.	Nyctaginaceae	Herb	He	Forb	+	+	+	+
45.	*Bombax ceiba* L.	Malvaceae	Tree	Ph		+	+	+	+
46.	*Borreria articularis* (L.f.)F.N.Williams	Rubiaceae	Herb	Th	Forb	–	+	–	–
47.	*Bothriospermum tenellum* Hornem.	Boraginaceae	Herb	Th	Forb	+	–	–	–
48.	*Bougainvillea glabra* Willd.	Nyctaginaceae	Tree	Ph		+	+	+	+
49.	*Broussonetia papyrifera* Vent.	Moraceae	Tree	Ph		+	+	+	+
50.	*Butea monosperma* (Lamk).Thub.	Fabaceae	Tree	Ph		+	+	+	+

Contd...

Table 2.3–Contd...

S.No.	Name of Plant Species	Family	Habit	Life Form	Herb Category	Ju*	Gu*	Jo*	Aa*
51.	*Callicarpa macrophylla* Vahl	Verbenaceae	Shrub	Ph		+	+	–	+
52.	*Callistemon citrinus* (Curtis) Skeels	Myrtaceae	Tree	Ph		+	+	+	+
53.	*Calotropis procera* R.Br.	Asclepiadaceae	Shrub	Ch		+	+	+	+
54.	*Cannabis sativa* L.	Urticaceae	Shrub	Ch		+	+	+	+
55.	*Cassia fistula* L.	Caesalpiniaceae	Tree	Ph		+	+	+	+
56.	*Cassia mimosoides* L.	Caesalpiniaceae	Herb	Th	Legume	+	–	–	–
57.	*Cassia obtusifolia* L.	Caesalpiniaceae	Herb	Th	Legume	+	+	+	+
58.	*Cassia occidentalis* Vahl	Caesalpiniaceae	Herb	Th	Legume	+	+	+	+
59.	*Cathranthus roseaus* (L.) G.Don	Apocynaceae	Shrub	Ch		+	+	+	+
60.	*Cedrela toona* Roxb.	Meliaceae	Tree	Ph		+	+	+	+
61.	*Celosia argentea* L.	Amaranthaceae	Herb	Th	Forb	–	–	+	+
62.	*Celosia cristata* L.	Amaranthaceae	Herb	Th	Forb	+	+	+	+
63.	*Centella asiatica* (L.) Urb.	Apiaceae	Herb	He	Forb	+	+	+	+
64.	*Centipeda minima* L.	Asteraceae	Herb	Th	Forb	+	–	–	–
65.	*Ceratopteris thalictroides* Brong.	Parkiaseae	Fern	G		+	–	–	–
66.	*Cestrum nocturnum* L.	Solanaceae	Shrub	Ph		+	–	–	–

Contd...

Table 2.3–Contd...

S.No.	Name of Plant Species	Family	Habit	Life Form	Herb Category	Ju*	Gu*	Jo*	Aa*
67.	*Chloris dolichostachya* Lag.	Poaceae	Herb	Th	Grass	+	+	+	+
68.	*Chrysopogon aciculatus* (Retz.) Trin.	Poaceae	Herb	Th	Grass	–	+	–	–
69.	*Chrysopogon montanus* Spreng.	Poaceae	Herb	Th	Grass	+	+	+	+
70.	*Chrysopogon serrulatus* Trin.	Poaceae	Herb	Th	Grass	+	+	+	+
71.	*Cissampelos pareira* L.	Menispermaceae	Herb	Th	Forb	+	+	+	+
72.	*Cissus repanda* Vahl	Vitaceae	Herb	Th	Forb	–	–	+	–
73.	*Citrus aurantifolia* Christm.	Rutaceae	Tree	Ph		+	–	–	–
74.	*Cheilanthes albomarginata* C.B.Clarke	Sinopteridaceae	Fern	G		–	–	+	–
75.	*Clematis gounana* Roxb.	Ranunculaceae	Climber	Th		–	–	–	+
76.	*Cleome viscosa* DC.	Capparaceae	Herb	Th	Forb	–	–	–	+
77.	*Clerodendrum viscosum* Vert.	Verbenaceae	Shrub	Th		+	+	+	+
78.	*Coccinia cordifolia* Cogn.	Cucurbitaceae	Herb	Th	Forb	+	+	+	+
79.	*Colebrookia oppositifolia* Smith	Lamiaceae	Shrub	Ph		+	+	–	+
80.	*Colocasia esculanta* (L.) Schott.	Araceae	Shrub	G		+	+	+	+
81.	*Commelina benghalensis* L.	Commelinaceae	Herb	Th	Forb	+	+	+	+
82.	*Commelina padulosa* Blume	Commelinaceae	Herb	Th	Forb	+	+	+	+
83.	*Conyza stricta* Willd.	Asteraceae	Herb	Th	Forb	+	+	–	–

Contd...

Table 2.3–Contd...

S.No.	Name of Plant Species	Family	Habit	Life Form	Herb Category	Ju*	Gu*	Jo*	Aa*
84.	*Corchorus aestuans* L.	Tiliaceae	Herb	Th	Forb	+	+	+	+
85.	*Cordia myxa* L.	Boraginaceae	Tree	Ph		+	+	+	+
86.	*Coronopus didymus* L.	Cruciferae	Herb	Th	Forb	+	+	–	–
87.	*Crotalaria medicaginea* Lamk.	Fabaceae	Herb	Th	Legume	+	+	+	+
88.	*Crotalaria mucronata* Desv.	Fabaceae	Herb	Th	Legume	+	–	–	+
89.	*Crotalaria spectobilis* Roth	Fabaceae	Herb	Th	Legume	+	–	–	–
90.	*Crotalaria tetragona* Roxb.	Fabaceae	Herb	Th	Legume	+	+	–	+
91.	*Croton sparsiflorum* Morong.	Euphorbiaceae	Herb	Th	Forb	+	–	–	–
92.	*Crytococcum accrescens* (Trin.) Stapf	Poaceae	Herb	Th	Grass	+	+	–	–
93.	*Cucumis hardwickii* Royle.	Cucurbitaceae	Herb	Th	Forb	–	+	+	–
94.	*Curculigo orchioides* Gaertn.	Hypoxidaceae	Herb	G	Forb	+	+	+	+
95.	*Cuscuta reflexa* Roxb.	Convolvulaceae	Climber	Ph		+	+	+	+
96.	*Cyanotis cristata* (L.) D.Don	Commelinaceae	Herb	Th	Forb	+	+	–	–
97.	*Cyanotis vaga* L.	Commelinaceae	Herb	Th	Forb	+	+	–	–
98.	*Cyathocline purpurea* D.Don	Asteraceae	Herb	Th	Forb	–	–	–	+
99.	*Cynodon dactylon* (D) Pers.	Poaceae	Herb	He	Grass	+	+	+	+

Contd...

Table 2.3–Contd...

S.No.	Name of Plant Species	Family	Habit	Life Form	Herb Category	Ju*	Gu*	Jo*	Aa*
100.	*Cynoglossum lanceolatum* Forssk.	Boraginaceae	Herb	Th	Forb	–	+	+	–
101.	*Cynoglossum zeylanicum* Vahl. ex. Horn.	Boraginaceae	Herb	Th	Forb	+	+	+	+
102.	*Cyperus alulatus* Kern	Cyperaceae	Herb	Th	Sedge	+	+	–	+
103.	*Cyperus brevifolius* (Rottf.) Hassk.	Cyperaceae	Herb	Th	Sedge	+	+	+	+
104.	*Cyperus iria* L.	Cyperaceae	Herb	Th	Sedge	+	+	–	–
105.	*Cyperus niveus* Retz.	Cyperaceae	Herb	Th	Sedge	+	+	+	+
106.	*Cyperus paniceus* (Rottb.) Boeck.	Cyperaceae	Herb	Th	Sedge	+	+	–	–
107.	*Cyperus pilosus* Vahl	Cyperaceae	Herb	Th	Sedge	+	+	+	+
108.	*Cyperus pumilus* L.	Cyperaceae	Herb	Th	Sedge	+	+	+	+
109.	*Cyperus rotundus* L.	Cyperaceae	Herb	G	Sedge	+	+	+	+
110.	*Cyperus squarrosus* (L.) Cent.	Cyperaceae	Herb	Th	Sedge	+	–	–	–
111.	*Dactyloctenium aegyptiacum* Beauv.	Poaceae	Herb	He	Grass	+	+	+	+
112.	*Dalbergia sisso* Roxb.	Fabaceae	Tree	Ph		+	+	+	+
113.	*Datura stromonium* L.	Solanaceae	Herb	Ch	Forb	+	+	–	–
114.	*Delonix regia* Raf.	Caesalpiniaceae	Tree	Ph		+	+	+	+
115.	*Dendrocalamus strictus* Nees.	Poaceae	Tree	Ph		+	+	–	+

Contd...

Table 2.3–Contd...

S.No.	Name of Plant Species	Family	Habit	Life Form	Herb Category	Ju*	Gu*	Jo*	Aa*
116.	*Desmodium cephalotes* Wall.	Fabaceae	Herb	Th	Legume	+	+	–	–
117.	*Desmodium concinnum* DC.	Fabaceae	Herb	Th	Legume	+	+	+	+
118.	*Desmodium gangiticum* (L.) DC.	Fabaceae	Herb	Th	Legume	+	+	+	+
119.	*Desmodium pulchellum* Benth. ex Baker	Fabaceae	Herb	Th	Legume	+	+	+	+
120.	*Desmodium velutinum* (Willd.) DC.	Fabaceae	Herb	Th	Legume	+	+	+	+
121.	*Desmostachya bipinnate* Stapf	Poaceae	Herb	He	Grass	+	+	+	+
122.	*Dicliptera roxburghiana* Nees	Acanthaceae	Herb	He	Forb	+	+	+	+
123.	*Digitaria cruciata* Nees	Poaceae	Herb	Th	Grass	+	–	+	+
124.	*Digitaria granuloris* Trin. ex Spreng.	Poaceae	Herb	Th	Grass	+	+	+	+
125.	*Digitaria setigera* Roth.	Poaceae	Herb	Th	Grass	+	+	+	+
126.	*Digitaria stricta* Roth. ex R. & S. Syst.	Poaceae	Herb	Th	Grass	+	+	+	+
127.	*Dioscorea bellophylla* (Prain). Voigt ex Haines.	Dioscoreaceae	Herb	G	Forb	+	+	+	+
128.	*Dioscorea bulbifera* L.	Dioscoreaceae	Herb	G	Forb	+	+	–	+
129.	*Dioscorea pentaphylla* L.	Dioscoreaceae	Herb	G	Forb	+	+	–	–
130.	*Diplazium esculentum* (Retz.) Sac.	Athyriaceae	Fern	G		+	+	+	+
131.	*Dolichos biflorus* L.	Fabaceae	Herb	Th	Legume	–	+	–	–

Contd...

Table 2.3–Contd...

S.No.	Name of Plant Species	Family	Habit	Life Form	Herb Category	Ju*	Gu*	Jo*	Aa*
132.	*Echinochloa frumentacea* L.	Poaceae	Herb	Th	Grass	–	+	–	+
133.	*Eclipta prostrata* Roxb.	Asteraceae	Herb	Th	Forb	+	+	+	+
134.	*Ehretia laevis* Roxb.	Boraginaceae	Tree	Ph		+	+	–	+
135.	*Elephantopus scaber* L.	Asteraceae	Herb	Th	Forb	+	+	+	+
136.	*Eleusine indica* Gaertn.	Poaceae	Herb	Th	Grass	+	+	–	–
137.	*Emblica officinalis* Gaertn.	Euphorbiaceae	Tree	Ph		–	+	–	–
138.	*Emilia sonchifolia* DC.	Asteraceae	Herb	Th	Forb	+	+	–	+
139.	*Equisetum diffusum* D.Don	Equisetaceae	Fern	G		+	+	–	–
140.	*Eragrostis atrovirens* (Desf.) Trin ex Steud.	Poaceae	Herb	He	Grass	+	+	–	+
141.	*Eragrostis japonica* (Thunb.) Trin.	Poaceae	Herb	He	Grass	+	+	+	+
142.	*Eragrostis tenella* Roem. & Schult.	Poaceae	Herb	He	Grass	+	+	+	+
143.	*Eragrostis unioloides* (Retz.) Nees ex Steud.	Poaceae	Herb	He	Grass	+	+	+	+
144.	*Erianthus ravennae* L. P. Beauv.	Poaceae	Herb	He	Grass	+	+	–	–
145.	*Erigeron bonariensis* L.	Asteraceae	Herb	Th	Forb	+	+	+	+
146.	*Eucalyptus hybrid* L.Herit.	Myrtaceae	Tree	Ph		+	+	+	+
147.	*Eugenia jambolana* Lam.	Myrtaceae	Tree	Ph		+	+	+	+

Contd...

Table 2.3–Contd...

S.No.	Name of Plant Species	Family	Habit	Life Form	Herb Category	Ju*	Gu*	Jo*	Aa*
148.	*Euphorbia helioscopia* L.	Euphorbiaceae	Herb	Th	Forb	+	+	–	–
149.	*Euphorbia hirta* L.	Euphorbiaceae	Herb	Th	Forb	+	+	+	+
150.	*Euphorbia hypericifolia* L.	Euphorbiaceae	Herb	Th	Forb	–	+	–	–
151.	*Euphorbia pulcherrima* Willd. ex Klotz.	Euphorbiaceae	Herb	Th	Forb	–	+	+	–
152.	*Evolvulus nummularious* L.	Convolvulaceae	Herb	Th	Forb	–	+	–	+
153.	*Fagopyrum cymosum* Meissn.	Polygonaceae	Herb	Th	Forb	–	–	–	+
154.	*Ficus bengalensis* L.	Moraceae	Tree	Ph		–	+	–	–
155.	*Ficus infectoria* Roxb.	Moraceae	Tree	Ph		+	+	+	+
156.	*Ficus palmata* Forssk.	Moraceae	Tree	Ph		+	–	–	–
157.	*Ficus racemosa* L.	Moraceae	Tree	Ph		+	+	+	+
158.	*Ficus religiosa* L.	Moraceae	Tree	Ph		+	+	+	+
159.	*Ficus roxburghii* Wall.	Moraceae	Tree	Ph		–	+	–	–
160.	*Ficus scandens* Roxb.	Moraceae	Tree	Ph		+	+	–	–
161.	*Fimbristylis dichotoma* (L.) Vahl	Cyperaceae	Herb	Th	Sedge	+	+	+	+
162.	*Fimbristylis melicea* (L.) Vahl. Enum.	Cyperaceae	Herb	Th	Sedge	+	+	+	–
163.	*Fumaria indica* Haussk.	Fumariaceae	Herb	Th	Forb	+	+	–	–

Contd...

Table 2.3–Contd...

S.No.	Name of Plant Species	Family	Habit	Life Form	Herb Category	Ju*	Gu*	Jo*	Aa*
164.	*Galium vestitum* D.Don	Rubiaceae	Herb	Th	Forb	+	–	–	–
165.	*Glycosmis pentaphylla* Corrca.	Rutaceae	Shurb	Ph		+	+	+	+
166.	*Gnaphalium luteoalbum* L.	Asteraceae	Herb	Th	Forb	+	+	+	+
167.	*Gomphrena celosioides* Mart.	Amaranthaceae	Herb	Th	Forb	+	+	+	+
168.	*Goodyera biflora* Hook.	Orchidaceae	Herb	Th	Forb	+	–	–	–
169.	*Gossypium herbaceum* L.	Malvaceae	Shurb	Ph		+	+	–	+
170.	*Grevillea robusta* A. Cunn. ex R.Br.	Proteaceae	Tree	Ph		–	+	–	–
171.	*Grewia oppositifolia* Roxb.	Tiliaceae	Tree	Ph		+	+	–	–
172.	*Gymnostemma pentaphyllum* (Thunb.) Ma Kino	Cucurbitaceae	Herb	Th	Forb	–	–	+	–
173.	*Haplophagma adenophyllum* Wall.	Bignoniaceae	Tree	Ph		+	–	–	+
174.	*Hedyotis pinifolia* Wall.	Rubiaceae	Herb	Th	Forb	–	+	–	–
175.	*Helicteres isora* L.	Sterculiaceae	Shrub	Ph		+	+	+	–
176.	*Helminthostachys zeylanica* L.	Ophioglossaceae	Fern	G		+	–	+	–
177.	*Hemigraphis rupestris* (Heyne ex T. Andr.)	Acanthaceae	Herb	Th	Forb	+	+	+	+
178.	*Heteropogon contortus* L.	Poaceae	Herb	He	Grass	+	+	–	–
179.	*Hibiscus lobatus* (J.A.Murr.) O.Ktze	Malvaceae	Shrub	Th		+	–	–	–

Contd...

Table 2.3–Contd...

S.No.	Name of Plant Species	Family	Habit	Life Form	Herb Category	Ju*	Gu*	Jo*	Aa*
180.	*Hibiscus rosa-sinensis* L.	Malvaceae	Shrub	Ph		+	+	–	–
181.	*Holarrhena antidysenterica* Wall.	Apocynaceae	Tree	Ph		+	+	+	+
182.	*Holoptelea integrifolia* Planch.	Ulmaceae	Tree	Ph		+	+	+	+
183.	*Houttuynia cordata* Thunb.	Piperaceae	Herb	Th	Forb	–	+	–	+
184.	*Hydrocotyl javanica* Thunb.	Apiaceae	Herb	He	Forb	+	–	–	–
185.	*Hygrophila auriculata* Schum.	Acanthaceae	Herb	Th	Forb	+	–	–	–
186.	*Ichinocarpus fructescens* R.Br.	Apocynaceae	Climber	Th		+	+	+	+
187.	*Imperata cylindrica* L.	Poaceae	Herb	He	Grass	+	+	–	–
188.	*Indigofera tinctoria* L.	Fabaceae	Herb	Th	Legume	–	–	+	–
189.	*Ipomoea eriocarpa* R.Br.	Convolvulaceae	Herb	Th	Forb	+	+	+	+
190.	*Ipomoea hedrifolia* L.	Convolvulaceae	Herb	Th	Forb	+	+	–	+
191.	*Ipomoea nil* L.	Convolvulaceae	Herb	Th	Forb	+	+	+	+
192.	*Ipomoea plamata* Forssk.	Convolvulaceae	Herb	Th	Forb	–	–	–	+
193.	*ipomoea quamoclit* L.	Convolvulaceae	Herb	Th	Forb	+	+	+	+
194.	*Isachne miliacea* Roth.	Poaceae	Herb	Th	Grass	+	+	–	–
195.	*Jatropha curcas* L.	Euphorbiaceae	Tree	Ph		–	+	–	–

Contd...

Table 2.3–Contd...

S.No.	Name of Plant Species	Family	Habit	Life Form	Herb Category	Ju*	Gu*	Jo*	Aa*
196.	*Justicia adhatoda* Nees	Acanthaceae	Shrub	Ch		–	–	+	–
197.	*Justicia procumbens* L.	Acanthaceae	Herb	Th	Forb	+	+	+	+
198.	*Justicia simplex* D.Don	Acanthaceae	Herb	Th	Forb	–	+	–	–
199.	*Knoxia sumatrensis* Retz.	Rubiaceae	Herb	Th	Forb	+	+	+	+
200.	*Kydia calycina* L.	Malvaceae	Tree	Ph		+	+	+	+
201.	*Lactuca sativa* L.	Asteraceae	Herb	Th	Forb	–	+	–	–
202.	*Lannea coromandelica* Hontt.	Anacardiaceae	Tree	Ph		+	+	+	+
203.	*Lantana camara* L.	Verbenaceae	Shrub	Ph		+	+	+	+
204.	*Lantana indica* Roxb.	Verbenaceae	Shrub	Ph		–	+	+	–
205.	*Launaea aspleniifolia* Hook.f.	Asteraceae	Herb	Th	Forb	+	+	–	–
206.	*Leea asiatica* L.	Leeaceae	Shrub	Th		–	+	–	–
207.	*Leonotis nepetaefolia* R.Br.	Lamiaceae	Herb	Th	Forb	+	+	–	+
208.	*Lepidagathis caspidata* Ness	Acanthaceae	Herb	Th	Forb	+	+	+	+
209.	*Lepidagathis incurva* (Buch.Ham)	Acanthaceae	Herb	Th	Forb	–	–	+	+
210.	*Lepidagathis purpuricaulis* Nees	Acanthaceae	Herb	Th	Forb	+	+	+	+
211.	*Leptadenia reticulata* W & A.	Asclepiadaceae	Herb	Th	Forb	–	–	–	+

Contd...

Table 2.3–Contd...

S.No.	Name of Plant Species	Family	Habit	Life Form	Herb Category	Ju*	Gu*	Jo*	Aa*
212.	*Lespedeza gerardiana* Garh. ex Baker	Fabaceae	Herb	Th	Legume	+	+	–	–
213.	*Leucas cephalotes* (Roth) Spreng.	Lamiaceae	Herb	Th	Forb	+	+	+	+
214.	*Lichi chinensis* Sonner.	Sapindaceae	Tree	Ph		–	–	–	+
215.	*Limnophila rugosa* Roth.	Scrophulariaceae	Herb	Th	Forb	+	+	–	+
216.	*Lindenbergia indica* L.	Scrophulariaceae	Herb	Th	Forb	+	–	–	–
217.	*Lindernia anagallis* (Burm.f.)	Scrophulariaceae	Herb	Th	Forb	+	+	+	+
218.	*Lindernia antipoda* L.	Scrophulariaceae	Herb	Th	Forb	–	+	–	–
219.	*Lindernia ciliata* Colsm.	Scrophulariaceae	Herb	Th	Forb	+	+	+	+
220.	*Lindernia parviflora* L.	Scrophulariaceae	Herb	Th	Forb	+	–	–	–
221.	*Phyla nodiflora* (L.) Greene	Verbenaceae	Herb	Th	Forb	+	+	–	+
222.	*Ludwigia octovalvis* (Jacq.) Raven.	Onagraceae	Herb	Th	Forb	+	+	+	+
223.	*Ludwigia prostrata* Roxb.	Onagraceae	Herb	Th	Forb	+	–	–	–
224.	*Lygodium flexuosum* (L.) Sw.	Lygodiaceae	Fern	G		+	–	+	–
225.	*Mallotus philippenensis* Muell. Arg.	Euphorbiaceae	Tree	Ph		+	+	+	+
226.	*Malvastrum coromandelianum* Garcke.	Malvaceae	Herb	Th	Forb	+	+	+	+
227.	*Mangifera indica* L.	Anacardiaceae	Tree	Ph		+	+	+	+

Contd...

Table 2.3–Contd...

S.No.	Name of Plant Species	Family	Habit	Life Form	Herb Category	Ju*	Gu*	Jo*	Aa*
228.	*Mazus japonicus* Thunb.	Scrophulariaceae	Herb	Th	Forb	+	+	–	–
229.	*Mazus pumilus* Burm.f.	Scrophulariaceae	Herb	Th	Forb	+	+	+	+
230.	*Melia azedarach* L.	Meliaceae	Tree	Ph		+	+	+	+
231.	*Melilotus indica* All.	Fabaceae	Herb	Th	Legume	+	+	+	+
232.	*Melochia corcorifolia* L.	Sterculiaceae	Herb	Th	Forb	+	+	+	+
233.	*Mentha sylvestris* L.	Lamiaceae	Herb	Th	Forb	+	–	–	–
234.	*Mimosa pudica* L.	Mimosaceae	Shrub	Ch		+	+	+	+
235.	*Mitrogyna parvifolia* Korth.	Rubiaceae	Tree	Ph		+	+	–	–
236.	*Mollugo pentaphylla* L.	Molluginaceae	Herb	Th	Forb	–	–	–	+
237.	*Morus alba* L.	Moraceae	Tree	Ph		+	+	+	+
238.	*Mukia moderaspatana* L.	Cucurbitaceae	Herb	Th	Forb	+	–	–	–
239.	*Murdannia nudiflora* (L.) Brenan	Commelinaceae	Herb	Th	Forb	+	+	+	+
240.	*Murdannia spirata* (L.) Bruckn.	Commelinaceae	Herb	Th	Forb	–	–	–	+
241.	*Murraya koenigii* Spreng.	Rutaceae	Shrub	Ph		+	+	+	+
242.	*Nelumbo speciosum* Willd.	Nymphaeaceae	Shrub	G		+	–	–	+
243.	*Nepeta hindostana* Roth.	Lamiaceae	Herb	Th	Forb	–	+	–	–

Contd...

Table 2.3–Contd...

S.No.	Name of Plant Species	Family	Habit	Life Form	Herb Category	Ju*	Gu*	Jo*	Aa*
244.	*Nerium indicum* Mill.	Apocynaceae	Tree	Ph		+	–	–	–
245.	*Nicotiana plumbaginifolia* Viv.	Solanaceae	Herb	Th	Forb	+	+	–	–
246.	*Nyctanthes arbortristis* L.	Oleaceae	Tree	Ph		–	+	–	–
247.	*Ocimum americanum* L.	Lamiaceae	Herb	Th	Forb	+	–	–	–
248.	*Oldenlandia diffusa* Willd.	Rubiaceae	Herb	Th	Forb	+	+	–	–
249.	*Oldenlandia offinis* (Roen. & Schwt.) DC.	Rubiaceae	Herb	Th	Forb	+	+	–	+
250.	*Ophioglossum reticulatum* L.	Ophioglossaceae	Fern	G		+	–	–	–
251.	*Oplismenus burmannie* (Retz.) P. Beauv. Ess. Agrost.	Poaceae	Herb	Th	Grass	+	+	+	+
252.	*Oplismenus compositus* Beauv.	Poaceae	Herb	Th	Grass	+	+	+	+
253.	*Opuntia dillenii* Haw.	Cactaceae	Shrub	Ph		–	+	–	–
254.	*Orthosiphon rubicundus* (D.Don) Benth	Lamiaceae	Herb	Th	Forb	+	+	+	+
255.	*Oxalis acetosella* L.	Oxalidaceae	Herb	He	Forb	+	+	+	+
256.	*Oxalis corniculata* L.	Oxalidaceae	Herb	He	Forb	+	+	+	+
257.	*Oxalis dehradunensis* Raizada	Oxalidaceae	Herb	He	Forb	+	–	–	–
258.	*Parthenium hysterophorus* L.	Asteraceae	Herb	Th	Forb	+	+	+	+
259.	*Paspalidium flavidum* Retz.	Poaceae	Herb	Th	Grass	+	+	+	+

Contd...

Table 2.3–Contd...

S.No.	Name of Plant Species	Family	Habit	Life Form	Herb Category	Ju*	Gu*	Jo*	Aa*
260.	*Paspalum paspalodes* (Michx.) Scribn.	Poaceae	Herb	Th	Grass	+	+	+	+
261.	*Perilla frutescens* (L.)Britt.	Lamiaceae	Herb	Th	Forb	–	+	–	+
262.	*Peristrophe bicalyculata* Retz.	Acanthaceae	Herb	Th	Forb	+	+	+	+
263.	*Perotis hordeliformis* Nees. ex Willd.	Poaceae	Herb	Th	Grass	–	+	–	–
264.	*Peucedanum dhana* Ham.	Apiaceae	Herb	Th	Forb	+	–	–	–
265.	*Phalaris minor* Retz.	Poaceae	Herb	Th	Grass	+	+	+	+
266.	*Phoenix acaulis* Buch-Ham.	Arecaceae	Tree	Ph		+	–	–	+
267.	*Phoenix sylvestris* Roxb.	Arecaceae	Tree	Ph		+	–	–	+
268.	*Phyllanthus debilis*	Euphorbiaceae	Herb	Th	Forb	+	–	–	–
269.	*Phyllanthus fraternus* Webster.	Euphorbiaceae	Herb	Th	Forb	+	+	–	–
270.	*Phyllanthus niruri* L.	Euphorbiaceae	Herb	Th	Forb	+	+	+	+
271.	*Phyllanthus reticulatus* Poir.	Euphorbiaceae	Shrub	Ph		+	–	–	+
272.	*Physalis minima* L.	Solanaceae	Herb	Th	Forb	+	+	+	+
273.	*Piper nepalense* Miq. (E.)	Piperaceae	Herb	Th	Forb	+	+	–	–
274.	*Plantago exiqua* L.	Plantaginaceae	Herb	Th	Forb	–	+	–	–
275.	*Plectranthus japonicus* Burm.f.	Lamiaceae	Herb	Th	Forb	+	+	–	+

Contd...

Table 2.3–Contd...

S.No.	Name of Plant Species	Family	Habit	Life Form	Herb Category	Ju*	Gu*	Jo*	Aa*
276.	*Plumbago zeylanica* L.	Plumbaginaceae	Shrub	Ch		+	–	–	–
277.	*Pogostemone benghalense* (Burm.f.) Kuntz	Lamiaceae	Shrub	Th		+	+	+	+
278.	*Polyalthia longifolia* Benth & Hook. F.	Annonaceae	Tree	Ph		+	+	+	+
279.	*Polygonum barbatum* L.	Polygonaceae	Herb	Th	Forb	+	+	+	+
280.	*Polygonum hydropiper* L.	Polygonaceae	Herb	Th	Forb	+	+	+	+
281.	*Polygonum plebejum* R.Br.	Polygonaceae	Herb	Th	Forb	+	+	–	+
282.	*Polypogon fugax* Nees ex Steud	Poaceae	Herb	Th	Grass	+	+	+	+
283.	*Populus alba* L.	Salicaceae	Tree	Ph		+	+	+	+
284.	*Porana paniculata* Roxb.	Convolvulaceae	Herb	Th	Forb	–	+	–	–
285.	*Portulaca oleracea* L.	Portulacaceae	Herb	Th	Forb	+	+	+	+
286.	*Pothus cathcarti* Schott.	Araceae	Climber	G		–	–	+	–
287.	*Pouzolzia indica* Gaud.	Urticaceae	Herb	Th	Forb	+	+	+	+
288.	*Pouzolzia zeylanica* Benn.	Urticaceae	Herb	Th	Forb	–	+	+	–
289.	*Prunus persica* Batsch.	Rosaceae	Tree	Ph		–	+	–	–
290.	*Pteris vitata* L.	Pteridaceae	Fern	G		+	+	+	+
291.	*Pupalia lappacea* (L).Juss.	Amaranthaceae	Herb	Th	Forb	–	+	–	–

Contd...

Table 2.3–Contd...

S.No.	Name of Plant Species	Family	Habit	Life Form	Herb Category	Ju*	Gu*	Jo*	Aa*
292.	*Ranunculus scleratus* L.	Ranunculaceae	Herb	Th	Forb	+	+	+	+
293.	*Rauwolfia serpentina* Benth.	Apocynaceae	Herb	Th	Forb	–	+	+	–
294.	*Ricinus communis* L.	Euphorbiaceae	Shrub	Ph		+	+	+	+
295.	*Rottbaellia exaltata* L.	Poaceae	Herb	Th	Grass	+	+	–	+
296.	*Rubus niveus or Rubus lasiocarpus* Wall.	Rosaceae	Climber	Ch		+	–	–	–
297.	*Rumex dentatus* L.	Polygonaceae	Herb	Th	Forb	+	+	+	+
298.	*Rungia pectinata* L	Acanthaceae	Herb	Th	Forb	+	+	+	+
299.	*Saccharum spontaneum* L.	Poaceae	Herb	He	Grass	+	+	+	+
300.	*Salvia plebeia R.Br.*	Lamiaceae	Herb	Th	Forb	+	+	+	+
301.	*Saussurea heteromalla* D.Don	Asteraceae	Herb	Th	Forb	+	+	+	+
302.	*Schleichera oleaosa* Laur.	Sapindaceae	Tree	Ph		+	+	–	+
303.	*Scoparia dulcis* L.	Scrophulariaceae	Herb	Th	Forb	+	+	+	+
304.	*Selaginella subdiaphana* (Wall. ex Hook & Grev.) Spring	Selaginellaceae	Fern	G		+	+	–	+
305.	*Sesamum orientale* L.	Pedaliaceae	Herb	Th	Forb	+	+	+	+
306.	*Setaria glauca* Beauv.	Poaceae	Herb	Th	Grass	+	+	+	+
307.	*Setaria homonyma* (Steud.) Choiv.	Poaceae	Herb	Th	Grass	+	+	+	+

Contd...

Table 2.3–Contd...

S.No.	Name of Plant Species	Family	Habit	Life Form	Herb Category	Ju*	Gu*	Jo*	Aa*
308.	*Setaria paniculifera* (Steud.) Foura ex Heml.	Poaceae	Herb	Th	Grass	+	+	–	–
309.	*Setaria tomentosa* Roxb.	Poaceae	Herb	Th	Grass	+	+	–	–
310.	*Sida acuta* Burm.	Malvaceae	Herb	Th	Forb	+	+	+	+
311.	*Sida cordifolia* L.	Malvaceae	Herb	Th	Forb	+	+	+	+
312.	*Siegesbeckia orientalis* L.	Asteraceae	Herb	Th	Forb	+	+	–	–
313.	*Solanum incanum* L.	Solanaceae	Shrub	Ph		+	+	+	–
314.	*Solanum nigrum* L.	Solanaceae	Herb	Th	Forb	+	+	+	+
315.	*Solanum surattense* Burm.	Solanaceae	Shrub	Ph		+	+	–	+
316.	*Solanum verbascifolium* L.	Solanaceae	Shrub	Ph		+	+	–	+
317.	*Solena heterophylla* Lour.	Cucurbitaceae	Herb	G	Forb	+	–	–	–
318.	*Solvia anthemifolia* (Juss.) R.Br.	Asteraceae	Herb	Th	Forb	+	+	–	–
319.	*Sonchus oleraceous* L.	Asteraceae	Herb	Th	Forb	+	+	+	+
320.	*Sphenoclea zeylanica* Gaertn.	Sphenocleaceae	Herb	Th	Forb	–	+	–	–
321.	*Sporobolus diander* Beauv.	Poaceae	Herb	Th	Grass	+	+	+	+
322.	*Stellaria media* L.	Caryophyllaceae	Herb	Th	Forb	+	+	+	+
323.	*Stephania glabra* Roxb.	Menispermaceae	Herb	Th	Forb	–	–	+	–
324.	*Tabermontana divaricata* L.	Apocynaceae	Shrub	Ph		–	+	–	–

Contd...

Table 2.3–Contd...

S.No.	*Name of Plant Species*	*Family*	*Habit*	*Life Form*	*Herb Category*	*Ju**	*Gu**	*Jo**	*Aa**
325.	*Tectona grandis* L.f.	Verbenaceae	Tree	Ph		+	+	+	+
326.	*Tephrosia purpurea* Pers.	Fabaceae	Herb	Th	Legume	+	+	+	+
327.	*Terminalia arjuna* W & A.	Combretaceae	Tree	Ph		+	+	–	–
328.	*Terminalia bellerica* Roxb.	Combretaceae	Tree	Ph		–	+	–	–
329.	*Terminalia chebula* Retz.	Combretaceae	Tree	Ph		–	+	–	–
330.	*Terminalia tomentosa* W & A.	Combretaceae	Tree	Ph		+	–	–	+
331.	*Thelypteris dentata* (Forssk.) St. John	Thelypteridaceae	Ferm	G		+	+	–	+
332.	*Thelypteris prolifera* Retz.	Thelypteridaceae	Fern	G		+	+	+	+
333.	*Themeda arundinacea* (Roxb.) Ridley	Poaceae	Shrub	He		+	+	–	+
334.	*Thevetia peruviana* (Pers.) Schum	Apocynaceae	Shrub	Ph		–	+	–	–
335.	*Tinospora cordifolia* Miers.	Menispermaceae	Climber	Th		+	+	–	–
336.	*Tinospora sinensis* (Lour.) Merr.	Menispermaceae	Climber	Th		–	+	+	+
337.	*Torenia cordifolia* Roxb.	Scrophulariaceae	Herb	Th	Forb	+	+	+	+
338.	*Trewia nudiflora* L.	Euphorbiaceae	Tree	Ph		+	+	+	+
339.	*Trichosanthes cordata* Roxb.	Cucurbitaceae	Herb	Th	Forb	–	+	+	–
340.	*Tridax procumbens* L.	Asteraceae	Herb	Th	Forb	+	+	+	+
341.	*Triumfetta pentandra* A.Rich.	Tiliaceae	Herb	Th	Forb	+	+	+	+

Contd...

Table 2.3–Contd...

S.No.	Name of Plant Species	Family	Habit	Life Form	Herb Category	Ju*	Gu*	Jo*	Aa*
342.	*Triumfetta rhomboidea* Jacq. Enum.	Tiliaceae	Herb	Th	Forb	+	+	+	+
343.	*Typha elephantina* Roxb.	Typhaceae	Shrub	He		+	+	+	+
344.	*Urena lobata* L.	Malvaceae	Herb	Th	Forb	+	+	+	+
345.	*Ventilago caliculata* Tulasne.	Rhamnaceae	Climber	Ph		–	–	–	+
346.	*Vernonia cinerea* Less.	Asteraceae	Herb	Th	Forb	+	+	+	+
347.	*Vernonica anagallis-aquatica* L.	Scrophulariaceae	Herb	Th	Forb	+	+	+	+
348.	*Vetiveria zizaniodes* Nash.	Poaceae	Herb	He	Grass	+	+	+	+
349.	*Vicia sativa* L.	Fabaceae	Herb	Th	Legume	+	+	+	+
350.	*Vitex negundo* L.	Verbenaceae	Shrub	Ph		–	+	–	+
351.	*Wrightia tomentosa* Roem & Sch.	Apocynaceae	Tree	Ph		–	+	–	–
352.	*Xanthium strumarium* L.	Asteraceae	Shrub	Ch		+	+	+	+
353.	*Youngia japonica* DC.	Asteraceae	Herb	Th	Forb	+	+	–	+
354.	*Zingiber capitatum* Roxb.	Zingiberaceae	Herb	G	Forb	+	+	+	+
355.	*Zizyphus jujuba* Lamk.	Rhamnaceae	Shrub	Ph		+	+	+	+
356.	*Zizyphus xylopyra* Willd.	Rhamnaceae	Shrub	Ph		+	+	+	+
	Total					290	291	201	236

Abbreviations: +: Present; –: Absent; Ph: Phanerophyte; Ch: Chamaephyte; He: Hemicryptophyte; G: Geophyte; Th: Therophyte.

* Ju: Jurkha; Gu: Gulzarpur; Jo: Jogipura; Aa: Aampani.

species, and minimum 201 species were recorded in Jogipura forest. The habit approach revealed that herbs contributed more than 50 per cent of plant species followed by trees, shrubs and climbers. In the study sites, the most of the herbaceous plants and trees were obtained in mixed deciduous forests (Jurkha and Gulzarpur) hile the least number of herbaceous plants and trees was recorded in Jogipura forest because it suffered from heavy anthropogenic pressure. Lower values (minimum number) might be because of the damage done to the plants by gazing/browsing (Weaver and Clements, 1966).

Table 2.4: Various Plant Life Forms and their Composition in the Vegetation Community

Life Form	*Total No.*	*Percentage of the Total*
Trees	60	16.85
Shrubs	39	10.96
Climbers	9	2.52
Ferns	14	3.93
Herbs	234	65.73
Total	356	100.00
Herb category		
Grasses	38	16.23
Sedges	11	4.70
Legumes	21	8.97
Forbs	164	70.08
Total herbs	234	100.00

Applying Raunkiaer's spectra, the flora was therocryptophytic because of dominance of therophytes and cryptophytes in the forests. Because forests of Tarai and Bhabar are affected by deforestation, the percentage of phanerophytes is lower than that of therophytes in all study sites. The study sites are situated in Tarai and Bhabar of Kumaun and near the banks of Kosi and Dabka rivers; so various small grazing lands also situated in these area. Local villagers depend on forage for their domestic animals in these forests. High value for therophytes is an indicator of the amount of influence of man and animals (Bharucha and Dave, 1944). According to Cain

(1950), overgrazing tends to increase the percentage of therophytes through introduction and spread of weedy grasses. According to Danserau (1957) the predominance of therophytes indicates warm climate.

Daubenmire (1968) attributed the occurrence of therophytes to aridity factor. Meher Homji (1964) concluded that the life forms of different regions of the country are reflected by the bioclimate of the area. Pandeya (1964), Yadav and Singh (1977) and Agrawal (1984) attributed the preponderances of therophytes to the influence of heavy grazing or biotic interferences. Saxena *et al.* (1982) concluded that along the major altitudinal gradient (from tropical to alpine), the phanerophytes decrease in importance, hemicryptophytes increase in importance, chamaephytes, geophytes and therophytes do not show much alternation. According to Pandit (2003), increase in the percentage of therophytes and decrease in that of phanerophytes and hemi cryptophytes in the flora of Gir forest was caused by deforestation, intensive utilization of land for cultivation and grazing by livestock.

On the basis of above study, it is concluded that a system of education must be developed by government to make the local people of the area aware of the importance of conserving the vegetation. Excess influence of local villagers in forests should be prohibited. There should be programme under which local villagers can be educated about forest and its natural products. The area needs attention so that the vegetation could be restored.

Acknowledgement

One of us (BJ) gratefully thanks Prof. Y.P.S. Pangtey, D.Sc. F.N.A.Sc., and Dr. Lalit Tewari of Kumaun University, Nainital for their help, encouragement, moral support, providing necessary suggestions and motivation during this study.

References

Adhikari, B.S., Joshi, M., Rikhari, H.C. and Rawat, Y.S. 1992. Cluster Analysis (Dendrogram) of high altitude (2150-2500 m) forest vegetation around Pindari glacier in Kumaun Himalaya. *Journal of Environmental Biology* 13: 101-105.

Adhikari, B.S., Rikhari, H.C., Rawat, Y.S. and Singh, S.P. 1991. High altitude forests composition, diversity and profile structures in a part of Kumaun Himalaya. *Tropical Ecology* 32: 86-97.

Agrawal, A.K. 1990. Floristic composition and phenology of temperate grasslands of Western Himalaya as affected by scraping, fire and heavy grazing. *Plant Ecology* 88(2): 177-187.

Agrawal, K.B. 1984. Plant, Climate, Species diversity and concentration of dominance in some grazing lands of Garhwal Himalayas. *Geobios* 11: 168-173.

Bankoti, N.S. 1990. Woody vegetation analysis along elevational gradient (2000-3600 m) of Pindari catchment (Kumaun Himalaya). Ph.D. thesis, Kumaun University, Nainital, India.

Beg, A.R. 1975. *Wildlife Habitats of Pakistan*. Bull. No. 5. Pakistan Forest Institute, Peshawar.

Bharucha, F.R. and Dave, R.N. 1944. The biological spectrum of grassland association. *J. University Bombay* 13: 15-16.

Cain, S.A. 1950. Life-forms and phytoclimate. *Bot. Rev.* 16: 1-32.

Champion, H.G. and Seth, S.K. 1965. Forest types of Pakistan. Pakistan Forest Institute, Peshawar. 233pp.

Chapin, F.S. III, Walker, B.H., Hobbs, R.J., Hopper, D.U., Lawton, J.H., Sola, D.E., and Tilman, D. 1997. Biotic control over the functioning of ecosystem. *Science* 277: 500-504.

Collet, H. 1971. *Flora Simlensis*. 3rd impression.

Danserau, P. 1957. Description and recording of vegetation upon a structural basis, *Ecology* 32: 172-229.

Daubenmire, R.F. 1968. Plant Communities, A textbook of Plant Synecology. Harper and Row, New York, 300 pp.

Dhar, U. 2000. Prioritization of conservation sites in the timberline zone of West Himalaya. *In*: Singh, S., Sastry, A.R.K., Mehta, R. & Uppal, V. (eds) Setting Biodiversity Conservation Priorities for India Vol I. WWF India, pp 193-211.

Dhar, U., Rawal, R.S. and Samant, S.S. 1997. Structural diversity and representativeness of forest vegetation in a protected area of Kumaun Himalaya. India. *Biodiversity and conservation* 6: 1045-1062.

Gupta, R.K. 1968. *Flora Nainitalensis*. Navayug Traders. New Delhi. 489 pp.

Jain, S.K. and A.R.K. Sastry, A.R.K. 1983. Botany of some tiger habitats in India. *Botanical Survey of India, Howrah.* pp. 5-8.

Kala, C.P. and Uniyal, V.K. 1999. Forest vegetation along an altitudinal gradient in the Valley of flower National Park and its vicinity, Western Himalayan. *Annals of Forestry* 7: 60-69.

Kumari, V. and Biswas, S. 2003. Forest biodiversity conservation study of Indian Himalayan eco-sensitive sites.

Meher- Homji, V.M. 1964. Life forms and biological spectrum as enharmonic criteria of acidity and humidity in Trops. *J. Indian. Bot. Soc.* 43(3): 423-430.

Mooney, H.A., Cushman, J.H., Medina, E. Sala, O.E. and Schulze, E.D. 1996. *Functional Role of Biodiversity: A Global Perspective.* John Wiley and Sons, Chichster, U.K.

Osmaston, A.E. 1926. *A Forest Flora for Kumaun.* International Book Distributors, Dehradun, India.

Pandeya, S.C. 1964. Ecology of grass land of Sagar (M. P.) I.I.A. Composition of the fenced grassland association. *J. Indian Bot. Soc.* 43: 577-605.

Pandit. B.R. 2003. Biodiversity of Tropical dry deciduous forest ecosystem. XII World Forestry congress 2003, Quebec City, Canada.

Pant, P.C. 1976. Plants of Corbett National Park. Uttar Pradesh. *Journal of the Bombay National History Society* 78: 287-295.

Pant, P.C., Uniyal, B.P. and Prasad, R. 1981. Additions to the plants of Corbett National Park. Uttar Pradesh. *Journal of the Bombay Natural History Society* 78: 50-3.

Raizada, M.B. 1978. *Flora of Mussoorie.* Bishen Singh & Mahendra Pal Singh, Dehradun. 645 pp.

Ralhan, P.K., Saxena, A.K and Singh, J.S. 1982. Analysis of forest vegetation at and around Nainital in Kumaun Himalaya. *Proceedings of Indian National Science Academy* 48(B): 121-137.

Raunkiaer, C. 1934. *The life forms of plants and statistical geography.* Claredon, Oxford, 632p.

Rawal, R.S., Bankoti, N.S. and Pangtey, Y.P.S. 1994. Broad community identification of high altitude forest vegetation in Pindari region of Kumaun (Central Himalaya). *Proceedings of Indian National Science Academy* B60: 553-556.

Rikhari, H.C., Adhikari, B.S. and Rawat, Y.S. 1997. Woody species composition of temperate forest along an elevational gradient in Indian Central Himalaya. *J. Trop. For. Sci.* 10: 197-211.

Rikhari, H.C., Ralhan, P.K. and Singh, S.P. 1989. Phytosociology and population structure of chir-pine forests in Kumaun Himalaya. *Annals of Biology* 5: 129-140.

Samant, S.S. 1999. Diversity, nativity, and endemism of vascular plants in a part of Nanda Devi Biosphere reserve in West Himalaya I. *Himalayan Biosphere Reserve (Biannual Bulletin)* 1: 1-28.

Saxena, A.K., Pandey, P. Singh, J.S. 1982. Biological spectrum and other structural functional attributes of the vegetation of Kumaun Himalaya. *Plant Ecology* Vol. 49: 111-119.

Saxena, A.K., Pandey, T. and Singh, J.S. 1985. Altitudinal variation in the vegetation of Kumaun Himalayas. pp. 43-66. *In:* D.N. Rao. K.J. Ahmed, M. Yunus & S.N. Singh (eds.) *Perspectives in Environmental Botany*. Print House. Lucknow.

Silori, C.S. 2001. Status and distribution of anthropogenic pressure in the buffer zone of Nanda Devi Biosphere Reserve in western Himalaya, India. *Biodiversity and Conservation* 10(7): 1113-1130.

Singh, J.S. and Singh, S.P. 1984. *An Integrated Ecological Study of Eastern Kumaun Himalaya with Emphasis on Natural Resources*. Vol. 1-3, Final Report (HCS/DST/187/76). Kumaun University, Nainital.

Singh, J.S. and Singh, S.P. 1987. Forest vegetation of the Himalaya. *The Botanical Review* 53(1): 80-192.

Singh, J.S. and Singh, S.P. 1992. *Forests of Himalaya: Structure, Functioning and Impact of Man*. Gyanodaya Prakashan, Nainital, India.

Singh, J.S., Pandey, U. and Tiwari, A.K. 1984. Man and Forests: A Central Himalayan Case Study. *Ambio* 13: 80-87.

Singh, J.S.and Singh, S.P. 1987. Structure and functioning of central Himalayan chir-pine forest ecosystem. *Current Science* 56: 383-391.

Singh, R.S., Ralhan, P.K. and Singh, S.P. 1987. Phytosociological and population structure of mixed Oak conifer forest in a part of Kumaun Himalaya. *Environmental Ecology* 5: 475-487.

Tewari, J.C. and Singh, S.P. 1981. Vegetation analysis of a forest lying in transitional zone between lower and upper Himalayan moist temperate forests. pp. 104-119. *In*: G.S.Paliwal (ed.). *The Vegetational Wealth of Himalaya*. Puja Publishers, New Delhi.

Tewari, J.C. and Singh, S.P. 1985. Analysis of woody vegetation in mixed Oak forest of Kumaun Himalaya. *Proceedings of Indian National Science Academy* 51: 332-347.

Upreti, N., Tewari, J.C. and Singh, S.P. 1985. The oak forests of Kumaun Himalaya (India). Composition, diversity and regeneration: *Mountain Research and Development* 5: 163-174.

Vitousek, P.M., Mooney, H.A., Lubchenko, J. and Melillo, J.M. 1997. Human domination of Earth's ecosystem. *Science* 277: 500-504.

Weaver, J.E. and Clements, F.E. 1966. *Plant Ecology*. TMH Edition (2), New Delhi. pp. 1-601.

Yadav, P.S. and Singh, J.S. 1977. Grassland vegetation; its structure, function, utilization and management. Today and Tomorrows. *Printer and Publishers, New Delhi*, 132 pp.

Chapter 3

Introduced Species: A Threat to Biodiversity

☆ *M.K. Srivastava*

So far as the improvement in the existing life forms on the earth is concerned, it's a well known fact that the chance of improvement is directly proportional to the available genetic base and vice-versa. Loss of habitat, changing in climate pattern, entry of new alien species etc. are the prime cause of loss in biodiversity. New crop species (Bt crops) are also a threat to our natural flora and fauna as the alien gene has the potentiality to get reshuffle with indigenous base. The extinction of species and destruction of ecosystems has been a major concern of policy makers. To realize the importance of the matter, Government has established crop wise National Gene Banks that are provided with the facility of Seed Repository, Cryo-Bank and tissue culture repository. Pre- Breeding is also a useful approach to enhance genetic base in crop plants. National parks and sanctuaries have been established in many countries to conserve life forms in their natural way.

Introduction

Biodiversity includes the diversity of species, life forms, ecosystems and genes that exist on earth. The extinction of species and destruction of ecosystems has been a major concern of policy makers. One of the major efforts has been to conduct a survey and conserve country's biodiversity, so as to save wild plants and animals from extinction. National parks and sanctuaries have been established in many countries to meet this objective. At present we are very much listening about the new crop species (Bt crops, Genetically modified (GM) crops etc.) that have some alien gene (introduced from soil bacteria) and these genes have the potentiality to produce some toxic proteins in host plant. So the feeder on these crops may die because of toxic protein.

India is a developing country and 7 out of 10 people are dependent on agriculture. The land holding is marginal. As a botanist we know that surrounding crops may inter-cross and thus affect the neighboring crops. The pollens of Bt based crops can be blown and pollinate the surrounding non-Bt crops thus transferring the toxic gene to the non host crop too.

Another aspect of these genetically modified crops is that they may have herbicide tolerance. So when herbicide spray on the field except the crop in question remains withstand and rest vegetation (including weeds, some other flora and fauna) gets destroyed, in such a situation evolution of new weed species (Super weed) may be a reality (Bawa *et al.*, 2004) as seen in Georgia province of South America where a new weed species popularly known as PIGWEED evolved that cannot be controlled by any herbicide. So this is a burning question before the policy makers that how and how much they can allow the new crop species in our environment so that they can remain feasibly fit (James, 2006).

Reasons to Loss of Biodiversity

Biodiversity faces several threats. Some of the reasons are elaborated under the following heads:

Extinction

Extinction of species is one of the major characteristics of the post-industrial era. In many groups of plants and animals, there are group of species, which are prone to extinction due to environmental change so that the group is reduced to one or few threatened or endangered species. These threatened or endangered species have high probability of extinction during the next few years or decades (IUCN, 2007). Red Data books are prepared which contain the lists of these endangered species of plant and animals. The extinction of these rare species may be natural or manmade or both (http://www.molecular-plant-biotechnology.info/biotechnology-environments/biodiversity-conservation/extinction-and-endangered-species.html).

Natural Extinction

Natural extinction is primarily owing to small size of breeding population (thus promoting inbreeding and inbreeding depression) associated with high coefficient of variation. When the size of breeding population drops to a hundred or less, the likelihood of extinction is enhanced still further due to inbreeding depression.

Human Caused Extinction

Due to destruction of natural habitats by human interference, biodiversity is being lost at a fast rate, particularly in the tropical regions (*e.g.* tropical rain forest and coral reefs).

Parasites and Diseases

Bacteria, viruses, and parasites introduced with exotic species represent a threat to native organisms.

Food Chain Alteration

The feeding activity of exotics can alter the availability of food resources for native species.

Loss of Endemic Species

Competition for food and/or space may result in the elimination of related species occupying the same niche. Predation on a native species by exotics is also a problem.

Hybridization

Exotic species may reproduce with native populations, resulting in hybrids and the ultimate loss of the native species. In spite of these, there are other several reasons like urbanization, industrialization, disturbances with the ecosystem (climate change), narrowing genetic base due to continuous inbreeding and more and more use of chemical pesticides, herbicides, fertilizers etc. also play a vital role in depleting our biodiversity.

Steps to Conserve Biodiversity

1. No undisturbed land be used for development or urbanization;
2. Catalogues of genetic resources and national biological inventories be prepared so that the threatened and endangered species may be protected against extinction;
3. In poor nations, birth rate should be lowered and sustainable, high yielding agricultural systems be developed, so that preservation and sustainable exploitation of biodiversity go hand in hand;
4. Measures should be taken to reduce emission of greenhouse gases and ozone destroying compounds;
5. To promote "Conservation Biology" in-situ and ex-situ conservation techniques should be applied.

National Efforts to Conserve Biodiversity

In situ conservation

In situ conservation involves conservation of plants and animals in their native ecosystems or even in man made ecosystems where they naturally occur. This type of conservation applies only to wild flora and fauna and to the domesticated animals and plants, *e.g.* national parks, sanctuaries, nature reserves, natural monuments, etc.

Ex situ Conservation Efforts in India

Ex situ conservation involes the conservation of samples of genetic diversity away from their field habitats. This is achieved through the establishment of "Gene Banks" which includes Genetic Resources Centers, Zoos, Botanical Gardens, Culture collections etc. India has signed a joint project with United States on plant genetic resources in 1988 with an aim to establish national gene banks for various crops throughout the country. A gene bank must have the following facility to conserve the germ plasm for longer duration (Vasan, 2005).

Seed repository is one where the seeds are stored at -20 degree Celsius. In Cryo Bank the seeds are stored at -196 degree Farenhite.

Tissue Culture Repository (10-25 degree Celsius)

In vitro conservation techniques including tissue culture and cryo- preservation offer distinct advantages. In view of this, in 1986 a National Policy for Plant Tissue Culture repository was established at National banks for plant Genetic resources (NBPGR), new Delhi. At national level crop wise gene banks have been established throughout the country and they all are well supported by NBPGR, New Delhi. A few renowned national level gene banks are: Central Rice Research Institute, Cuttack; Directorate of Wheat Research, Karnal; Indian Grassland and Fodder Research Institute, Jhansi, etc.

References

Bawa, K.S., Das, A. Karanath, U. Krishnaswamy, J. and Rao, M. 2004. Ecosystem Profile: *Western Ghat and Srilanka Hotspot.* Western Ghat Regions, Critical Ecosystem Partnership Fund.

http://www.molecular-plant-biotechnology.info/biotechnology-environments/biodiversity-conservation/extinction-and-endangered-species.html

Vasan, S. 2005. In the name of law: legality, illegality and practice in Jharkhand Forests. *Economical and Political Weekly* 40(41):4447-4450.

James, C. 2006. *Global Status of Commercialized Biotech/GM Crops in 2005.*

IUCN. 2007. *The 2004 IUCN Red List of Threatened Species,* IUCN 2008.

Chapter 4

Climate Change and Diversity of Aquatic Avifauna in the Wetlands of Hassan District, Karnataka

☆ *B.S. Ravikumar*

The avifauna diversity and density in the wetland of Hassan district, Karnataka, India, was studied for a period of three years. Ponds of Hassan district inhabit several local and migratory bird species. Reduction in water retention in these ponds in summer, weed infestation; variations in food availability in different seasons, threat of predation, and other factors influencing breeding activity of birds affected the avifauna diversity in the study area. This habitat attracted 33 bird species belonging to 18 families, which are local and migratory aquatic birds, waders and others. Highest population was recorded in the family Ardeidae and Rallidae. Interestingly, the area was found congenial for certain resident migrants, *viz.* Storks, Ibises and Herons. Other prominent residents were Moorhens, Jacanas, and Cormorants. The visitors include

Ringed plovers, Wagtails and Pintails. It was evident that Purple moorhen, White breasted water hen, Pond heron have developed high tolerance to this highly fluctuating habitat and human activity. I hope that this study would provide a preliminary database for further research.

Introduction

Birds have fascinated human beings in various ways by their valuable services. They play a vital role in various wetland ecosystems; their diversity is an indication of congenial habitat for survival (Jayson and Mathew, 2002). The diversified vegetation of the wetlands of Malnad of Karnataka attracts large number of aquatic avifauna. However, in the recent past, the natural wetland patches and forest areas are depleting (Uttangi, 2001) at a rapid rate due to the expansion of rice fields and urbanization (Basavarajappa, 2005a,b).

Daniel *et al*. (1990), Jayson and Mathew (2002) and others have reported on various species of birds in different vegetation types. However, reports on aquatic birds occurring in Hassan district of central Karnataka are not available.

Study Area

Physiographically, the study area of Hassan district occupies a central position in the state of Karnataka, and lies between 12°31′ and 13°33′N and 75°33′ and 76°38′ E at an altitude of 913 m and its total geographical area is about 6,850 sq km. The topography of the district is comprised of the hilly and plateau regions. It begins at the base of the Western Ghats and is continuous into the gently rolling Deccan plateau. The district has highest peak Jenkal Betta (1,389 m from mean sea level) and the lowest peak (150 m from mean sea level). The Ghat region is drained to the West by a number of rivers into the Arabian Sea at Mangalore and plateau is drained by Krishna and Cauvery into Bay of Bengal. The dam at Gorur near the confluence of the Yagachi and Hemavathy rivers will generate a reservoir. The District shows a wide range of climatic conditions with a mean daily minimum of 14°C and mean daily maximum of 27°C. The plateau is dotted with numerous irrigation tanks often supporting an interesting aquatic flora and fauna. For the present study, the Hassan district was selected. The landscape consists of vast stretches of bushy vegetation and irrigated paddy fields and forest plants. The natural vegetation is typically that of scattered bushy scrub represented by *Acacia arabica, A. farmesiana, Eupatorium indica, Parthenum hysterophorus, Saccharum spontaneum, Ipomoea reptans, Lantana camara* and stray trees like *Eucalyptus citriodora, E. tereticorms, Pheonx vulgaris, Pongamia pinnata, Mangifera indica, Tamarindus indica, Cocos nucifera,Tectona grandis,Terminalia tomentosa, Eugenia jambolana* and other flora (Basavarajappa, 1998, 2002, 2005a; Kamath, 1991).

Four different sites were selected at random covering an area of 683 ha. They were: (i) Arasikere Doddakere dominated with rice fields; (ii) Janivara kere with wet grassland; (iii) Kattebiluguli lake with wetlands; and (iv) Kalasina kere with paddy and uncultivable grazing pasture. Avifauna of wetland-ecosystems of Malnad area of Karnataka located at the margins of sparsely distributed rice fields is a small seasonal canal. The study sites were differently located at 35 km each. In each site, observations were made fortnightly during 2009-10.

Methodology

The Variable Width Line Transect Method described by Burnham *et al.* (1980) was adopted. The transect covered 2km in all the four sites for the enumeration of water birds. Altogether 96 line transects were laid and birds were observed using a binocular (10x30) and identified with the help of field guides (Ali, 1996; Ali and Ripley, 1983; 1987; Sonobe and Usui, 1993; Woodcock, 1980). The flora of natural vegetation was identified as per Gambel (1967). The following formula was used for calculating Relative Dominance.

Relative Dominance = ni x 100/N; where, ni = number of individuals of the species; N = the total number of individuals of all the species seen during the study period. To find out the diversity of birds the commonness index method of Jayson and Mathew (2002) was used. The Commonness Index is the average frequency of sighting of a species in one sampling at a site. Further, Percent Occurrence and Relative Abundance were calculated as follows.

$$\%\ \text{Occurrence} = \frac{\text{Number of species of each family}}{\text{Total number of different species}} \times 100$$

$$\text{Relative Abundance} = \frac{\text{Number of individuals of the species}}{\text{Number of individuals of all species}} \times 100$$

To determine the significant differences of water birds and relative composition of different families, the Kruskal Wallis One Way Analysis of Variance and the Friedman Two-Way Analysis of Variance tests were used as per Saha (1992).

Results and Discussion

Details such as family, relative dominance and status of aquatic avifauna in the study area are as given in Table 4.1. Thirty three species of water birds belonging to 18 families were recorded. Of all, family Ardeidae (5 species) was relatively dominant (15.15 per cent). It represented 11.53 per cent of the total number of water bird species surviving under wetland conditions of Malnad region of Karnataka. The most common and abundant species of Ardeidae family were the Cattle Egret (*Bubulcus ibis*) and Little Egret (*Egretta garzetta*) followed by Indian Pond Heron or Paddy bird (*Ardeola grayii*), Large egret (*Cosmerodius aubus*), Grey heron (*Ardea cinerea*), Little Grebe (*Tachybaptus ruficollis*), Purple Moorehen (*Porphyrio porphyrio*) (Table 4.1).

Interestingly, the density of water birds varied significantly (X2= 12.86, P> 0.03) in different wetland-ecosystems of Malnad region of Karnataka. Six species of birds namely, Eurasian spoon bill (*Platalea leucorodia)*, Bar Headed Goose (*Anser Indicus*), Northern Pin tail (*Anas acuta*), Yellow wattled lapwing (*Vanellus malabaricus*), Yellow wattled lapwing (*Vanellus malabaricus*), Black-winged Stilt (*Himantopus himantopus*) and and River tern (*Sterna aurantia*) were uncommon (Table 4.1), Great Cormorant (*Phalacrocorax carbo*), Water Cock (*Gallinula chloropus*), White breasted waterhen (*Amouromis phoenicurus*), Lesser Pied Kingfisher (*Halcyon pileata*), Commons sand piper (*Actitis hypoleucos*), Large Pied Wagtail (*Motacilla maderaspatensis*) were rarely seen in this area. Thus, the relative composition of aquatic birds belonging to different families varied significantly (X2 = 15.5, P > 0.12). Habitats with varied vegetation influence the diversity of bird species (Jayson and Mathew, 2002). It is presumed that the native flora (*i.e.*, bushy scrub, stray trees and the paddy fields) might have extended comfortable shelter and foraging grounds for water birds.

Table 4.1: Aquatic Avifauna Found in the Wetland Ecosystem of Hassan District of Karnataka

Sl.No	*Family*	*Common Name*	*Scientific Name*	*Relative Abundance*	*Status*
1.	Podicipedidae	Little Greb	*Tachybaptus ruficollis*	12.86	R
2.	Phalacrocoracidae	Great cormorant	*Phalacrocorax carbo*	0.03	RM
		Little cormorant	*Phalacrocorax niger*	4.76	RM
3.	Pelecanidae	Spot billed Pelican	*Pelecana philippensis*	4.61	RM
4.	Ardiedae	Cattle Egret	*Bubulcus ibis*	2.09	RM
		Large egret	*Casmerodius aubus*	1.29	RM
		Little egret	*Egretta garzetta*	2.65	RM
		Grey Heron	*Ardea cinerea*	3.93	R
		Indian heron	*Ardeola grayii*	1.54	RM
5.	Jacanidae	Bronze winged Jacana	*Metopidius indicus*	1.23	R
		Pheasant tailed Jacana	*Hydrophasianus chirugus*	1.36	R
6.	Ciconidae	Painted stork	*Mycteria leucocephala*	3.31	R
7.	Threskiornithidae	Eurasian spoon bill	*Platalea leucorodia*	6.67	RM
		Ori. White ibis	*Threskiornis elanocephalus*	2.43	M
8.	Rallidae	Common Coot	*Fulcia atra*	23.67	R
		Purple Moorhen	*Porphyrio Porphyrio*	6.40	RM
		Water cock	*Gallinula chloropus*	0.33	R
		White breasted waterhen	*Amaurornis phoenicurus*	0.37	RM

Contd...

Table 4.1–Contd...

Sl.No	*Family*	*Common Name*	*Scientific Name*	*Relative Abundance*	*Status*
9.	Anatidae	Spot billed Duck	*Anas poecilorhyncha*	3.58	R
		Bar Headed Goose	*Anser Indicus*	1.70	RM
		Cotton teal	*Nettapus coromandelianus*	0.57	RM
		Northern Pin tail	*Anas acuta*	0.79	M
10.	Charadridae	Red wattled lapwing	*Vanellus indicus*	0.90	M
		Yellow wattled lapwing	*Vanellus malabaricus*	1.15	R
		Black winged Stit	*Himantopus himantopus*	5.96	R
11.	Laridae	River tern	*Sterna aurantia*	6.18	R
12.	Alcedinidae	Small bule King fisher	*Alcedo attihis*	0.45	R
13.	Dacelonidae	White breasted Kingfisher	*Halcyon smymensis*	0.19	R
14.	Cerylidae	Lesser Pied king fisher	*Halcyon pileata*	0.12	R
15.	Accepitridae	Brahmini Kite	*Haliastur Indus*	0.28	RM
16.	Scolopacidae	Common Sand Piper	*Actitis hypoleucos*	1.62	R
17.	Ploceidae	Streaked weaver bird	*Ploceus manyar*	2.06	R
18.	Motacillidae	Large Pied wag tail	*Motacilla maderaspatensis*	0.75	R

As these bird species are heterogeneous in their feeding habit (Ali and Ripley, 1983, 1987), the available fauna, *viz.*, crabs, snails, calms, worms, insect larvae and pupae in the paddy fields and in water bodies may constitute their feed. Wetlands are potential sources for plankton life. They play an important role in the cycle of changes (*i.e.*, elements of organic matter eaten by bacteria and protozoa and in turn consumed by the insect larvae, rotifers and crustaceans), which form the basis of food for water birds (Uttangi, 2001). By feeding on insect pests, their larvae and pupae, water birds control large amount of pest population in paddy fields. Further, they provide manure. Hence, the Ecological role played by water birds in wetland-ecosystems of Malnad area is unique of its kind.

Table 4.2: Aquatic Avifauna Representation in Families

Sl.No.	*Family*	*Percentage Occurrence*	*Relative Abundance*	*Status*
1.	Podicipedidae	3.03	12.86	R
2.	Phalacrocoracidae	6.06	4.79	RM
3.	Pelecanidae	3.03	4.60	RM
4.	Ardiedae	15.5	11.53	RM
	Jacanidae	6.06	2.60	R
6.	Ciconidae	3.03	3.32	R
7.	Threskiornithidae	6.06	3.11	RM
8.	Rallidae	12.12	30.78	R
9.	Anatidae	12.12	6.67	R
10.	Charadridae	9.09	8.03	M
11.	Laridae	3.03	6.18	R
12.	Alcedinidae	3.03	0.45	R
13.	Dacelonidae	3.03	0.19	R
14.	Cerylidae	3.03	0.12	R
15.	Accepitridae	3.03	0.29	RM
16.	Scolopacidae	3.03	1.63	R
17.	Ploceidae	3.03	2.06	R
18.	Motacillidae	3.03	0.76	R

R: Resident; M: Migratory; RM: Resident Migratory.

The resident migrant (*i.e.*, birds that breed in one part of the area in one season and move to other parts within the state or country in a different season) birds such as Painted Stork (*Mycteria leucocephala*), Black Ibis (*Pseudibis papillosa*), Oriental White Ibis (*Threskiornis melanocephalus*), and Northern Pintail (*Anas acuta*) were winter migrants in this region. The breeding activities of Black Ibis and Oriental White Ibis were observed at different localities in the wetlands. Interestingly Streaked Weaver birds were observed in breeding activities on *Typha angustata*. The area is enriched by small water canals, streams, water bodies and sparsely distributed tall stray trees *viz.Cocos nucifera, Mangifera indica Pheonx ulgaris, Tamarindus indica, Eucalyptus citriodora* and *E. tereticorms*. The bushy scrub constituted by *Lantana camara, Saccharum spontaneum, Ipomoea reptans* and others in the vicinity of paddy fields might have attracted the Oriental White Ibis and Black Ibis. Moreover, they get good food *i.e.*, insects, crabs, insect larvae, pupae and small fishes from this ecosystem.

Perhaps, all these conditions might have influenced these resident migrants to breed. To support this, few nests of the Black Ibis and Oriental White Ibis were seen on a few stray coconut trees near paddy fields. Thus, the wetland ecosystems have provided suitable habitats for the survival of these birds. However, in the recent years, the uncultivable grazing fields, bushy scrub areas and wetland patches have been cleared (Uttangi, 2001) and the stray trees removed for paddy cultivation (Basavarajappa, 1998, 2004). Weedicides and insecticides are ruthlessly used in paddy fields to control weeds and insect pests.

Many birds associated with wetlands are forced to leave the habitat as villagers cut the emergent vegetation and the number of nests have declined in this area (Basavarajappa, 2002, 2005a,b). The reason for decline of nests of other birds in general and water birds in particular is not known. This requires thorough scientific monitoring and documentation. Therefore, the preservation of wetland patches is essential to restore water birds.

Conclusion

The present study emphasizes the need to conduct a detailed study on the status of wetland ecosystems, and biology of water birds to have accurate information on the ecological role of birds associated with wetlands. The wetland ecosystems of Karnataka are diversified with various aquatic birds and 33 species of wetland-associated birds survive in this area. However, in the recent past, the wetland ecosystems in this area have deteriorated or destroyed due to encroachment for the expansion of rice fields and urbanization. This would eventually alter the natural vegetation on the banks of water canals, streams and in the uncultivable wetland patches making it unsuitable for birds to roost and nest. Unaltered natural patchy vegetation is essential in wetland ecosystems, which makes it very important to take steps to preserve wetlands' flora and avifauna for the health of biodiversity of Malnad areas.

Conservation Measures

Further, the involvement and participation of women working in self help groups need to conserve the bird diversity through educating local people. Bird poachers are to be punished through enforcing Wildlife Act. Protection and improvement of habitat is urgently needed. High usage of pesticides should be strictly banned.

Acknowledgements

I am grateful to University Grant Commission, New Delhi for sponsoring minor research project for the present study. I am thankful to Dr. A. Thimme gowda, Principal, AVK College for Women, Hassan. Thanks are also due to my colleagues and my students for their help during the field study.

References

Ali, S. 1996. *The Book of India Birds*. BNHS, Oxford University Press, Bombay.

Ali, S. and Ripley, S.D. 1983. *Handbook of the Birds of India andPakistan.* Oxford University Press, Oxford.

Ali, S. and Ripley, S.D. 1987. *Compact Handbook of the Birds of India, Pakistan, together with those of Bangladesh, Nepal, Bhutan and Sri Lanka.* Oxford University Press, New Delhi.

Basavarajappa, S. 2002. A preliminary report on the Indian Crested Peafowl in Channagiri area, Karnataka. *Newsletter for Birdwatchers* 42(2): 20-21.

Basavarajappa, S. (in press). Habitatic features, relative abundance, sex-ratio and population decline of Pavo cristatus in Channagiri area of Karnataka, India. In: Gupta, V.K. (ed.) *Prospectives in Animal Ecology and Reproduction*. Vol. III. Daya Publishing House, New Delhi.

Burnham, K.P., Anderson, D.R. and Laake, J.L. 1980. Estimation of density from line transect sampling of biological populations. *Wildlife Monograph* 72. Daniels, R.J.R., M. Hegde and Gadgil.

Conservation and Management. Daya Publishing House, Delhi.

Gambel, J.S. 1967. *Flora of the Presidency of Madras*. (2nd Edn) Vol. I & II. Botanical Survey of India, Calcutta.

Jayson, E.A. and Mathew, D.N. 2002. Structure and composition of two bird communities in the southern Western Ghats. *Journal of theBombay Natural History Society* 99(1):8-25.

Kamath, U.S. 1991. Shimoga District Gazetteer, Karnataka.

Philips, J. 1959. Succession, development, the climax and complex organisms. An analysis of concepts. Part 1 & 2. *Journal of Ecology* 22: 559-571.

Venkateshwarlu (eds.). *Trends in Wildlife Biodiversity*

Woodcock, M. 1980. *Collins Hand Guide to the Birds of the Indian Sub-continent*. St. James Palace, London.

Chapter 5

Global Climate Change and Freshwater Fisheries

☆ *Rajesh, R.N. Ram, R.S. Chuhan, Prabjeet Singh, C.P. Singh, and Prem Kumar*

Both types of freshwater systems – lotic and lentic – are bound to be significantly affected by global climate change. Water temperatures, decreased dissolved oxygen levels and increased toxicity of pollutants will be some of the general effects of global climate change. Quality of fish habitat could be affected by altered hydrologic regimes and increased groundwater temperatures of lotic systems. Eutrophication might be intensified and stratification will likely become more pronounced and stronger in the lentic systems. Food webs and habitat availability and quality might be altered. Since fish physiology is linked to temperature, and fish have evolved to cope with specific hydrologic regimes and habitat niches, their physiology and life histories will be affected by alterations induced by climate change. Not only this, even genetic change is also possible and is the only biological option for fish that are unable to migrate or acclimate. Artisanal, commercial, and recreational fisheries worldwide depend upon freshwater fishes.

Impacted fisheries may make it difficult for developing countries to meet their food demand, and developed countries may experience economic losses. As it strengthens over time, global climate change will become a more powerful stressor for fish living in natural or artificial systems. Furthermore, human response to climate change (*e.g.*, increased water diversion) will exacerbate its already-detrimental effects. Global change is not going to cease even if greenhouse gas emissions are curtailed significantly, which has been predicted through certain Models.

Introduction

One of the most important changes in the global environment which affects the existence of all living things is the climate change. The climate change globally and locally also changes the aquatic habitat which, in turn directly and indirectly affects the environment. The study of global climate change is burdened with uncertainty at each level of analysis: emissions scenarios, the response of the climate to this forcing, the linkage between global circulation and regional models, and the resulting impacts of these changes (Baede *et al.*, 2001). However, the evidence that our climate is changing more rapidly now than in periods in the past is becoming overwhelming. Following the Industrial Revolution, humans have increasingly relied on fossil fuels as an energy source. Currently, about 80 per cent of the world's power is generated from fossil fuels (Bolin *et al.*, 1986; ACIA, 2004) and this fossil fuel combustion does produce greenhouse gases that can affect the global climate (Bolin *et al.*, 1986).

Aquatic environment is extremely diverse and influenced by the climate changes to a large extent both locally and globally. The changes in the ocean and its life are well studied at different levels with all possible details compared to the freshwater habitats. The water, in which the aquatic organisms dwell, is much more important to all living organisms of the earth and so the influence by climate changes on water becomes very much relevant in the present situations.

The purpose of this review is to present current knowledge about climate change and its impact upon freshwater fish populations. Freshwater fisheries range from small subsistence efforts to large commercial endeavors, often under artificial conditions, where millions of pounds of fish are produced. Freshwater fisheries are typically based on one or more species from the salmon family (Salmonidae), the minnow, carp, and barb family (Cyprinidae), the cichlid family (Cichlidae), and multiple families within the catfishes (Siluriformes). Though we will focus on these families, we will occasionally illustrate key concepts using other freshwater fishes.

Alterations in Fish Habitat

Climate change will present fishes with new environments through a number of mechanisms. These will not necessarily operate independently, and the possibility of synergy or interactions between them is probable.

Effect of Increased Temperature

All freshwater fishes are exotherms that cannot regulate their body temperature through physiological means (Moyle and Cech, 2004) and whose body temperatures are virtually identical to their environmental temperatures. These fishes may thermo regulate behaviorally, by selecting thermally heterogeneous microhabitats, but they are constrained by the range of temperatures available in the environment. Because biochemical reaction rates vary as a function of body temperature, all aspects of an individual fish's physiology, including growth, reproduction, and activity are directly influenced by changes in temperature (Franklin *et al.*, 1995). Therefore, increasing global temperatures can affect individual fish by altering physiological functions such as thermal tolerance, growth, metabolism, food consumption, reproductive success, and the ability to maintain internal homeostasis in the face of a variable external environment. Temperature tolerance ranges are species-specific and include both stenothermal (narrow thermal range) species such as Arctic charr (*Salvelinus alpinus*), and eurythermal (wide tolerance range) species such as common carp (*Cyprinus carpio*). Fish populations that are faced with changing thermal regimes may increase or decrease in abundance, experience range expansions or contractions, or face extinction.

Climate Change Effect on Dissolved Oxygen

Biologically available (dissolved) oxygen (DO) is much less concentrated in water than in air (Moyle and Cech, 2004). Oxygen enters the water column through diffusion from the atmosphere (this is potentially facilitated via turbulence and mixing) and by photosynthetic production (Kalff, 2000; Stickney, 2000). Plant, animal, and microbial aerobic respiration may decrease DO in the water column, especially at night when photosynthesis stops. Dissolved oxygen concentrations of 5 mg/l or more are acceptable for most aquatic organisms (Stickney, 2000), and concentrations below 2–3 mg/l are considered hypoxic (Kalff, 2000).

Oxygen solubility in water has an inverse relationship with water temperature. For example, water at 0°C holds about 14.6 mg/l, but water at 25°C can only hold about 8.3 mg/l (Kalff, 2000). Because the aerobic metabolic rates of most cold-blooded aquatic organisms increase with temperature, an increase in temperature both decreases the DO supply (through reduced saturation concentrations relative to air) and increases the biological oxygen demand (BOD) (Kalff, 2000). Fishes exposed to elevated water temperatures (*e.g.*, from climate change) can face an "oxygen squeeze" where the decreased supply of oxygen cannot meet the increased demand.

Effect of Toxicity

The toxicity of common pollutants (*e.g.*, organophosphates and heavy metals) to fish generally increases at higher temperatures (Murty, 1986). These increases in toxicity may result from the increased production of bio activated free radicals that are more toxic than the parent compound or from the increased uptake of the original toxin (Murty, 1986). Studies on bioaccumulation have shown a positive correlation between temperature and the uptake of anthropogenic, and natural (Alcaraz and Espina, 1995) pollutants. This increased uptake is thought to result from increased gill ventilation rates at warmer temperatures. An increase in fish metabolism also facilitates a faster depuration of toxicants.

Effect of Climate Changes in Stream and Groundwater Temperatures

Increases in stream and groundwater temperatures will likely have more pronounced effects in temperate zones because groundwater temperatures in tropical streams are not significantly lower than summer stream temperatures. In temperate stream systems, salmonid genera including Salmo, Oncorhynchus, and Salvelinus and warm water fishes rely on groundwater discharges for summer thermal refugia, especially in lower latitude and lower elevation streams (Meisner, 1990).

A study of two headwater streams in the Humber and Rouge river systems in southern Ontario, Canada indicated that a 4.1 °C increase in summer air temperature would reduce thermally suitable brook trout habitat by 42 per cent and 30 per cent respectively (Meisner, 1990). These findings suggest that this pattern of habitat loss would also be experienced in other temperate and subtropical areas.

Eutrophication

The natural trophic state of an aquatic system is a function of volume, water residence time, and nutrient input from the surrounding watershed (Kalff, 2000). However, human activity can also alter the trophic status of aquatic systems through anthropogenic enrichment or nutrient depletion and climate change (Klapper, 1991). Most cases of eutrophication result from the input of excess nutrients from urban and agricultural runoff and from sewage discharge (Klapper 1991). However, increases in temperature can also augment the productivity of a body of water by increasing algal growth, bacterial metabolism, and nutrient cycling rates (Klapper, 1991). Lower stream flows could increase water residence times and reduce flushing of nutrients from lake systems, thereby increasing trophic status (Bertahas *et al.*, 2006). When coupled with the input of anthropogenic pollutants, climate change can accelerate the eutrophication process (Klapper, 1991) or delay recovery from anthropogenic eutrophication (Jankowski *et al.*, 2006). On the other hand, climate change could reduce anthropogenic eutrophication; increased strength and duration of stratification could lead to an increased sequestration of nutrients in the hypolimnion, where they are not biologically available to primary producers (Magnuson, 2002a). Likewise, an increase in stream flows and pollutant flushing rates could reduce eutrophication rates (Magnuson, 2002b).

Changes in trophic state often negatively affect fish communities through direct effects on macroinvertebrate prey and through effects on the algal community that support the zooplankton. The general result of eutrophication in temperate lakes appears to be the replacement of economically important species such as salmonids and centrarchids with smaller, less desirable species such as some cyprinids (and gizzard shad (*Dorosoma cepedianum*) (Persson *et al.*, 1991).

Water Temperature Effects on Limnology

In larger lakes, small annual temperature increases have a pronounced effect on the timing and strength of stratification in lotic systems. As global temperatures rise, stratification in temperate zone lakes will strengthen (Topping and Bond, 1988). The deeper portion of large lakes, the hypolimnion, contains water that is not directly heated by solar radiation (Kalff, 2000). At the same time, the warmer water in the relatively thin epilimnion is continually heated by the sun, and the density gradient between the two layers prevents mixing between them (Kalff, 2000). An increase in global temperatures will strengthen stratification because increased heating of the epilimnion will intensify the temperature and density gradients between the two compartments, making mixing more difficult. Climate change will also prolong stratification events by heating the epilimnion sufficiently to form a density gradient earlier in the year.

The epilimnion is exposed to the atmosphere and experiences turbulence-induced mixing of O_2 and sufficient light to stimulate algal photosynthesis. However, because algal growth requires nutrients such as NO_x, PO_4, and Mg^{+2}, the epilimnion is characterized by limited amounts of nutrients (Goldman and Horne, 1983). The hypolimnion does not receive oxygen from the atmosphere, and low light levels at depths below the compensation point prevent photosynthesis and algal use of nutrients, so it is characterized by a limited DO supply and a large store of nutrients (Goldman and Horne 1983). Lake mixing is essential for the movement of oxygen to the hypolimnion and nutrients to the epilimnion, where they can be incorporated into the food web (Klapper, 1991).

Water Loss from Lentic Systems

Global climate change will also affect aquatic systems through changes in the hydrologic cycle: evaporation, evapotranspiration, and precipitation. Although history has not linked increased evapotranspiration with lower water levels in temperate lakes, higher temperatures and insolation could increase current water loss rates from these systems because evaporation rates may outstrip input from increased precipitation (Allan *et al.*, 2005).

In tropical systems, evaporation and evapotranspiration often already exceed precipitation in the dry; it is also unknown if increased water loss to the atmosphere will be offset by rising precipitation rates. Evaporation rates are a driving factor in tropical lakes. Therefore, changes in their water chemistry are largely driven by inflows, evaporation and precipitation.

Small changes in water levels of lentic systems will likely have minimal impacts on freshwater pelagic fishes but could have more serious consequences for species with narrow bathymetric ranges. Similarly, temperate fishes that rely upon littoral vegetation for reproductive success often fail to produce strong year classes in drought years when lake levels are low (Moyle and Cech, 2004). Small changes in water levels will eliminate these crucial habitats, forcing these fishes to use areas devoid of their vital habitat structures. This change would likely lead to population declines and might set off a cascade of new interspecific interactions.

Effects of Temperature on Fish Physiology

A possible effect of global warming is that the water temperatures in specific ecosystems would increase to levels that are suboptimal or lethal for resident fishes with limited dispersal ability or for those living in systems that prevent migration. All fish have a thermal range bounded on the upper end by their critical thermal maxima (CTMax) and on the lower end by their critical thermal minima (CTMin) (Becker and Genoway, 1979). Although acclimation to higher-than-ambient or near-lethal temperatures allows fish to adjust these critical limits by a few degrees (Myrick and Cech, 2003). When exposed to temperatures within the sub lethal range (bounded by the upper and lower incipient lethal temperatures), fishes can still be affected, positively or negatively, by the environmental temperature. Most of these effects result from the close coupling of fish metabolic rates with water temperature.

Whether species increase or decrease in abundance, experience range expansions or contractions, or face extinction will depend on the following factors:

1. Whether it is stenothermal or eurythermal and what region it inhabits (arctic, subarctic, temperate, subtropical, or tropical).
2. The magnitude and rate of the thermal change in that ecosystem.

3. The genetic response of the population to new selective pressures.
4. Opportunities to move to areas with appropriate thermal conditions (narrow endemics in small lakes or in rivers with impassable barriers may not be able to move to avoid lethal conditions).
5. Changes in the abundance or distribution of sympatric species that may be prey, predators, or competitors for resources.
6. The availability of necessary physical resources such as habitat and water quality parameters.

Temperate Fishes

The reproductive success of temperate fishes will be affected by global warming. Low over winter temperatures (cold-tempering) are often essential for the spawning success of temperate fishes such as salmonids (Gerdaux, 1998), channel catfish (*Ictalurus punctatus)* and white sturgeon (*Acipenser transmontanus*). Given the need for "cold tempering" among many temperate species, increases in temperature, especially in the colder months, may reduce the reproductive output of commercially important fishes near the lower latitudinal bounds of their distributions. This will affect species in natural systems differently depending upon their reproductive strategy. Equilibrium species (large eggs, small clutches, late sexual maturity) are most vulnerable to environmental change, periodic species (late maturity, small eggs, large clutches) are less vulnerable, and opportunistic species (early maturity, small eggs, small clutches, frequent spawning) are the least vulnerable (Poff *et al.*, 2001).

Reproductive success of adults and recruitment of juveniles may also be affected by changing thermal regimes. Environmental dissolved oxygen levels must be high enough to support aerobic metabolism in fishes (Moyle and Cech, 2004). Most fishes can maintain adequate levels of oxygen uptake at DO concentrations above 5 mg/l. When concentrations drop below 5 mg/l, many species employ physiological and behavioral adaptations to maintain adequate rates of oxygen uptake, but as DO concentrations drop below 2–3 mg/l, these adaptations often prove insufficient. Temperate species that naturally occur in habitats where environmental hypoxia is commonplace, such as the crucian carp

(*Carassius carassius*) and Sacramento blackfish (*Orthodon microlepidotus*) have physiological adaptations that allow them to tolerate environmental hypoxia. For example, the alligator gar *Lepisosteus spatula* breathes air through a modified swim bladder. However, for most fishes, dissolved oxygen concentrations do not have to drop to very low levels before physiological functions are affected. Sublethal hypoxia (2–3 mg/l) decreases feeding activity (Stickney 2000), swimming performance, and fitness of emerging larvae Low DO levels have also been linked to immuno suppression in fish (Stickney, 2000). Fish exposed to hypoxic conditions are less tolerant of environmental stresses (*e.g.*, disease, crowding, food limitation, thermal stress, and natural and anthropogenic toxins) than fish in normal (air-saturated) water.

Decreased levels of dissolved oxygen can reduce the growth and reproductive success of individual fishes and prevent them from dealing as effectively with other environmental challenges. Therefore, if climate change results in reduced levels of dissolved oxygen, it is possible that individual fish could be smaller, less fecund, and less likely to survive.

Tropical Fishes

Current climate change scenarios predict small temperature increases for the tropics (IPCC, 2001), it seems possible that tropical fish will not suffer the effects of climate change as much as temperate or polar fishes. Unlike most temperate fishes, however, tropical species live in systems where diurnal water temperature fluctuations regularly approach their incipient upper lethal limits (United Nations Economic Commission for Asia and the Far East 1972). Though tropical fishes can currently endure these temperatures, a slight (1–2°C) increase in regional temperatures may cause the daily temperature maxima to exceed these limits, particularly for populations that currently exist in thermally marginal habitats (Roessig *et al.*, 2004).

Recent work by Chaterjee *et al.* (2004) on common carp and rohu (*Labeo rohita*), important eurythermal aquaculture species on the Indian subcontinent, showed increased tolerance to elevated temperatures following acclimation to water temperatures of 30 °C and 35 °C (Chatterjee *et al.*, 2004). This suggests that there may be physiological adjustments to a new climate by eurythermal species in low latitudes. However, for fish currently living in marginal habitats, the effects of global warming would be similar to those predicted for temperate systems because tropical fish exhibit similar physiological symptoms when subjected to elevated temperatures. For example, temperatures of 39–40 °C inhibited swimming ability in the tilapia (*Tilapia mossambica*) despite the fact that many other tilapia species can survive at temperatures up to 41–42 °C (Chervinski, 1982).

Effect of Climate Change on Aquatic Communities

Although other factors such as geomorphology and hydrologic regime also determine available fish habitat (Poff *et al.*, 2001), distributions of many temperate zone freshwater fish species are determined by temperature isoclines that are elevational and latitudinal in nature (Moyle *et al.*, 1982; Baltz *et al.*, 1987). Fishes are limited at the highest latitudes and altitudes of their range by cold temperatures, primarily because the growing seasons are too short to allow juveniles to attain sufficient size in their first summer to stave off over winter starvation. The warm water temperatures at the lowest latitudes and elevations of their range also become limiting factors (Reist *et al.*, 2006).

There may also be local movements within drainages where fish from lower elevations are able to move upstream into stream reaches that were previously too cold for them. These fish would compete for space in their "new" ecosystems, possibly at the expense of native species that share similar thermal requirements, or, in some cases, have lower thermal tolerances (Carveth *et al.*, 2006).

Parasite transmission depends on host condition, the presence of intermediate hosts necessary for the parasite life cycle, water quality, and temperature so climate change may alter host-parasite dynamics by changing transmission opportunities and host susceptibility (Marcogliese, 2001). Temperate and subarctic zones fishes may experience increased parasite loads due to increased transmission opportunity. In temperate and subarctic zones, over winter temperatures are a major limiting factor on the standing stock of parasites (Marcogliese, 2001), through both direct and indirect mechanisms, so warmer winters could allow the possibility of year round infection and multiple generations of parasites in 1 year (Marcogliese, 2001). Conversely, in some situations warmer winter temperatures could reduce the impact of disease and parasitism. In temperate zones with large annual temperature ranges, many opportunistic parasites infect fishes in the early spring, when they are still weakened from harsh winter conditions. Therefore, higher temperatures associated with milder winters may lower infection rates by decreasing over winter stress. Although intra-annual variability appears to have decreased in recent years (through elevated winter temperatures), the frequency of extreme weather has increased (Magnuson, 2002) .

Regardless of the type of system (tropical, subtropical, temperature, polar), it is important to note that range shifts will occur at the species level, but not necessarily at the community or ecosystem level. Fishes (and other organisms) in the same community may not share the same thermal optima and tolerance limits, and their competitive ability can be affected by temperature (Taniguchi and Nakano, 2000). Stenothermal species (*e.g.*, salmonids) are most likely to experience range shifts while eurythermal species (*e.g.*, common carp) may be capable of adapting to a new thermal regime. This raises the possibility that ecosystem components like food webs, interspecific competition, and host-parasite interactions will be altered. Therefore, aquatic communities will have to adjust to new biotic pressures in addition to altered abiotic conditions. Though the altered ecosystems will, eventually, achieve a new steady state, the forms this state will take, and the amount of time required to adjust, are completely unknown.

Possible Responses to New Challenges

Fish faced with a changing environment must adapt, migrate, or perish. Fish populations, especially large or widely distributed ones often have high heterogeneity that varies within their geographic range (Laikre *et al.*, 2005). For example, genetic loci in the European hake (*Merluccius merluccius*)) and the mummichog *Fundulus heteroclitus* (Smith *et al.*, 1998) vary along a latitudinal cline. The current rate of climate change rivals or surpasses all other rates recorded in the fossil record (Allan *et al.*, 2005), so it is likely that the changing climate will outstrip the ability of some fish populations to respond to new selective pressures. Historic and fossil records indicate that animals and plants tend to shift their range in response to climate change, indicating that some genetic traits are relatively intractable (Thomas, 2005). The ability of species to adapt will depend upon their genetic plasticity and degree of specialization. Specialists adapted to local conditions will be more likely to go extinct than generalists (Allan *et al.*, 2005).

Because global climate change will shift the ranges of temperate fishes towards higher latitudes, it could have potentially serious impacts on some stream fishes, fishes in geographically isolated environments, and fish in lentic systems. Migration towards the poles may be an option for some species, but fishes in east–west oriented stream systems, *e.g.*, the southwestern United States eastern Africa (Allan *et al.*, 2005), and New Zealand would be prevented from moving to higher latitudes by excessive distance and physiological barriers like the ocean. Though some fishes in streams with an east west orientation could migrate toward high-elevation headwaters in search of cooler temperatures, they would likely be faced with an entirely new set of environmental challenges such as smaller stream size and unfamiliar habitat (*e.g.*, higher gradient, larger substrate, changes in turbidity). Similarly, fishes forced to seek optimal temperatures in north–south oriented systems would probably also have to cope with a new physical environment.

Fishes in relatively closed systems such as lakes will also be affected with potentially detrimental results. Fisheries managers could alleviate this phenomenon by translocating "trapped" fishes into more suitable watersheds (Shuter *et al.*, 2002). The introduction of fish to new environments is not always successful and raises environmental, political, and ethical questions (Beeton, 2002). When various Colorado subspecies of cutthroat trout (*Oncorhynchus clarki*) were introduced into fishless or reclaimed waters, they experienced a success rate of less than 50 per cent .The introduction of fish into naive waters is not a decision to be made lightly because of the possible negative consequences for organisms already in that environment. History has taught us that introduction of non-indigenous species into natural ecosystems can have disastrous results (Courtenay, 1997). For example, the Nile perch (*Lates niloticus*) was introduced into Lake Victoria in a well-meaning effort to improve the fishery. The endemic cichlids had not evolved with a large, cursorial predator and were subsequently decimated.

Possible Solutions

The future impacts of climate change on fisheries and aquaculture are still poorly understood. The key to minimizing negative impacts and maximizing opportunities will be understanding and promoting the wide range of creative adaptive strategies – implemented by public institutions or the private sector – and their interactions with existing policy, legal and management frameworks. Addressing the potential complexities of climate change interactions and their possible scales of impact requires the mainstreaming of cross-sectoral responses into governance frameworks. Responses are likely to be more timely, relevant and effective where they are brought into the normal processes of development and engage people and agencies at all levels. This requires not only the recognition of climate-related vectors and processes, and their interaction with others, but also the availability of sufficient information for effective decision-making and approaches that engage the public and private sectors. The potential for the spatial displacement of aquatic resources and people as a result of climate change impacts and the impacts on transboundary resources requires that existing regional structures and processes be strengthened or given more specific focus.

Policy and legal mechanisms that address these issues will need to be developed or enhanced. Regional market and trading mechanisms are also likely to be more important in linking and buffering supply variability and maintaining sectoral value and investment. Although generally perceived as having only negative impacts, climate change may provide the sector with an additional positive impetus to move towards sustainability. For example, the resilience and adaptive capacities of aquatic-resource ecosystems, fisheries and aquaculture production systems, and of aquatic-resource dependent communities should be increased by applying existing good governance and management principles and approaches. Such approaches include the ecosystem approach to fisheries (EAF) and the ecosystem approach to aquaculture (EAA) – which includes practices of adaptive and precautionary management based on appropriate social, economic, political and institutional incentives.Similarly, improving the sector's fuel, energy and post-harvest efficiency would reduce its carbon footprint while bringing it closer to its sustainable development objectives.

Future Perspective

The continued provision of food and livelihood security from fisheries and aquaculture systems will require additional multiscale understanding of the impacts of climate change and of the interacting contributions of fisheries and aquaculture to food and livelihood security. Significant knowledge gaps exist in relation to the responses and adaptations of marine and freshwater resources and ecosystems to climate change, including critical thresholds and points of no return. There is also considerable uncertainty over the synergistic interactions between climate change and other stressors (*e.g.* water use, eutrophication, fishing, agriculture and the use of alternative energy). This means that planning for uncertainty will need to consider the increased possibility of unforeseen events. Nevertheless, examples of past management practices in response to climate variability and extreme events can provide useful lessons for the future, even though they will have to be placed in context of greater uncertainty. Better knowledge will be required about who is or will be vulnerable with respect to climate change, and food and livelihood security impacts, and about how this vulnerability arises and can be addressed. Better communication and application of what is known will be essential in knowledge building. Innovative approaches will be needed in order to target financial instruments and create effective incentives to promote adaptation and mitigation efforts.

At national and international levels, the public sector will have an important role in leveraging and integrating public-sector and private-sector investment, interacting through market mechanisms to meet sectoral aims for climate change responses and food and livelihood security. Many of these approaches are new and will need to be tested in the sector. At national level, climate change action plans are likely to build on the FAO Code of Conduct for Responsible Fisheries (CCRF) and related international plans of action (IPOAs), guidelines and other instruments, incorporated into appropriately linked policy and legal frameworks and management plans. Responses will need to employ integrated ecosystem-based approaches for the sector throughout the entire resource extraction, processing, supply and value chain. The future implications of climate change will intensify the justification for finding policy consensus to reform capture fisheries while respecting national-sector characteristics. In addition, sectoral trade and competition issues linked with climate change mitigation and adaptation activities are likely to become more important at the global level.

Therefore, fisheries sector representation in the pertinent policy and legal development processes is imperative.

References

Arctic Climate Impact Assessment (ACIA). 2004. *Impacts of a Warming Arctic: Arctic Climate Impact Assessment*. Cambridge University Press, Oxford, UK.

Aclaraz, G. and Espina, S. 1995.Acute toxicity of nitrite in juvenile grass carp modified by weight and temperature. *Bull Enviorn Contam Toxicil*. 55:473-478.

Allan, J.D., Palmer, M. and Poff, N.L. 2005. Climate change and freshwater ecosystems. In: Lovejoy TE, Hannah L (eds) *Climate Change and Biodiversity*. Yale University Press, New Haven, CT, pp 274–295.

Baltz, D.M., Vondracek B., Brown, L.R, Moyle, P.B. 1987. Influence of temperature on microhabitat choice by fishes in a California stream. *Trans Am Fish Soc* 116:12–20.

Baede, A.P.M., Ahlonsou, E., Ding, Y. and Schimel, D. 2001. The climate system: an overview. In: Houghton JT (ed.) *Climate Change 2001: The Scientific Basis*. Cambridge University Press, Cambridge, UK,pp.88-98.

Becker, C.D. and Genoway, R.G. 1979. Evaluation of the critical thermal maximum for determing tolerance of freshwater fish. *Envion Biol Fishes* 4:245-256.

Beeton, A.M. 2002. Large freshwater lakes: present state, trends, and future. *Environ Conserv* 29:29–38.

Bertahas, I., Dimltriou, E., Laschou, S. and Zacharias, I. 2006. Climate change and agricultural pollution effects on the trophic status of a Mediterranean lake. *Acta Hydroch Hydrob* 4:349–359.

Bolin B., Doos, B.R., Jager J. and Warrick, R.A. 1986. *The Greenhouse Effect, Climatic Change, and Ecosystems*. John Wiley and Sons, Chichester, UK.

Carvth, C.J., Widmer, A.M. and Bonar, S.A. 2006. Comparison of upper thermal tolerances of native and non native fish species in Arizona. *Trans Am Fisher Soc*. 135:1433-1440.

Chatterjee, N.A., Pal, K., Manush, S.M., Das, T. and Mukherje, C.S. 2004. Thermal tolerance and oxygen consumption of Labeo rohita and Cyprinus carpio early fingerlings acclimated to three different temperatures. *J Thermal Biol* 29:265–270.

Chervinski, J. 1982. environmental physiology of tilapias. In: Pullin R.S.V., Lowe, M.C., Connel, R.H. (eds). *Biology and Culture of Tilapias*. International Center for Living Aquatic Resources Management, Manila, Phillipines, pp 119-28.

Courtenay, W.J.J. 1997. Nonindigenious fishes.In:Simberloff D, Schmits D.C., Brown T.C. (eds).Strangers in paradise:impact and management of nonindigenous species in Florida.Island Press, Washington, DC, pp 109-122.

Franklin, C.E., Johnston, I.A., Crockford, T, Kamunde, C. 1995. Scaling of oxygen consumption of Lake Magadi tilapia, a fish living at 37°C. *J Fish Biol*. 46:829-834.

George, D.G. and Hewitt, D.P. 1998. The influence of year-to-year changes in position of the Atlantic Gulf Stream on the biomass of zooplankton in Windermere North Basin, U. K. In: George DG (ed.) *Management of Lakes and Reservoirs During Global Climate Change*. Kluwer Academic Publishers, Dodrecht, Netherlands, pp 223–244.

Gerdaux, D. 1998. Fluctuations in lake fisheries and global warming. In: George DG (ed.) *Management of Lakes and Reservoirs during Global Climate Change*. Kluwer Academic Publishers, Dodrecht, The Netherlands, pp 263–272.

Goldman, C.E. and Horne, A.J. 1983. *Limonology*. McGraw-Hill, Inc., New York.

Intergovernmental Panel on Climate Change (IPCC). 2001. Climate Change 2001: *The Scientific Basis*. Cambridge University Press, Oxford, UK.

Jankowski, T., Livingstone, D.M., Buhrer, H., Forster, R. and Niederhauser, P. 2006. Consequences of the 2003 European heat wave for lake temperature profiles, thermal stability, and hypolimnetic oxygen depletion: Implications for a warmer world. *Limnol Oceanogr* 51:815–819.

Jiongxin, X. 2005. The water fluxes of the Yellow River to the sea in the past 50 years, in response to climate change and human activities. Environ Manage 35:620–631.

Kalff, J. 2000. *Limnology*. Prentice Hall, Upper Saddle River, New Jersey.

Klapper, H. 1991. *Control of Eutrophication in Inland Waters*. Ellis Horword Ltd., West Sussex, UK.

Laikre, L, Palm, S and Ryman N. 2005. Genetic population structure of fishes: implications for coastal zone management. *Ambio* 34:111–119.

Magnuson, J.J., 2002a. Future of adapting to climate change and variability. In: McGinn NA (ed.) *Fisheries in a Changing Climate*. American Fisheries Society, Bethesda, MD, pp 283–287.

Magnuson, J.J. 2002b. Signals from ice cover trends and variability. In: McGinn NA (ed) *Fisheries in a Changing Climate*. American Fisheries Society, Bethesda, MD, pp 3–14.

Marcogliese, D.J. 2001. Implications of climate change for parasitism of animals in the aquatic environment. *Can J. Zool.* 79:1331-1352.

Meisner, J.D. 1990. Potential loss of thermal habitat forbrook trout due to climatic warming in two southern Ontario streams. *Trans Am Fisher Soc.*119:282-291.

Moyle, P.B. and Cech, J.J. 2004. *Fishes: an Introduction to Ichthyology*, 5th edn. Prentice Hall, Englewood Cliffs, NJ.

Murty, A.S. 1986. *Toxicity of Pesticides to Fish*. I.CRC Press, Boca Raton, FL.

Myrick C.A., Cech, J.J. 2000. Temperature influences on California rainbow trout physiological performance. *Fish Physiol Biochem* 22:245–254.

Persson, L., Diehl, S., Johannson, L., Andersson, G. and Hamrin, S.F. 1991. Shifts in fish communities along the productivity gradient of temperate lakes-patterns and the importance of size-structure interactions. *J Fish Biol.*38:281-293.

Poff, N.L., Angermeir, P.L., Cooper, S.D., Lake, P.S., Fausch, K.D., Winemiller, K.O., Mertes, L.A.K., Oswood, M.W, Reynolds, J. and Rahel, F.J. 2001. Fish diversity in streams and rivers.In:Chapian FSI, Sla OE, Huber-Sannwald E (eds). *Global Biodiversity in a Changing Environment: Scenarios for the 21st Century*. Springer, New York, pp351-368.

Reist J.D., Wrona, F.J., Prowse, T.D., Power, M., Dempson, J.B., Beamish, R.J., King, J.R., Carmichael, T.J. and Sawatzky, C.D. 2006. General effects of climate change on Arctic fishes and fish populations. *Ambio* 35:370–380.

Roessig, J.M., Woodley, C.M., Cech, J.J., Jr. and Hansen, L.J. 2004. Effects of global climate change on marine and estuarine fishes and fisheries. *Rev Fish Biol Fisher* 14:251–275.

Shuter, B.J., Carline, R.F., McGinn, N.A.2002. Overview of the symposium. In: McGinn NA (ed) *Fisheries in a Changing Climate.* American Fisheries Society,Bethesda,MD, pp 3–14.

Shuter, B.J., Minns, C.K. and Lester, N. 2002. Climate change, freshwater fish, and fisheries: case studies from Ontario and their use in assessing potential impacts. In: McGinn NA (ed) Fisheries in a changing climate. American Fisheries Society, Bethesda, MD, pp 77–88.

Stickney, R.R. 2000. *Encyclopedia of Aquaculture.* John Wiley and Sons, New York.

Taniguchi, Y. and Nakano, S. 2000. Condition-specific competition: implications for the altitudinal distribution of stream fishes. *Ecology* 81:2027–2039.

Topping, J.C. Jr and Bond, J.P. 1998. *The Potential Impact of Climate Change on Fisheries and Wildlife in North America.* The Climate Institute, U.S. Environmental protection Agency, pp 153-170.

Thomas C.D. 2005. Recent evolutionary effects of climate change. In: Lovejoy TE, Hannah L (eds.) *Climate Change and Biodiversity.* Yale University Press, New Haven, CT, pp 75–90.

Part II

Biodiversity Conservation: New Perspectives and Applications

Chapter 6

Biodiversity and Climate Change Mitigation: Assessment of Blackgram Cultivars Utilizing Available Genetic Diversity

☆ *Piyusha Singh, Dipti Arya, Nandita Limbu and C.S. Kar*

The major constraints in achieving higher yield of blackgram are lack of genetic variability, absence of suitable genotypes for different cropping systems, poor harvest index and susceptibility to diseases. Therefore, there is need to develop improved blackgram cultivators utilizing available genetic diversity. The present study was conducted to investigate morphological differentiation among a number of selections derived from pedigree method of breeding in urdbean. Efforts were also made to elucidate variations among lines derived from advance segregating generations (F5 and F6) of same or different cross combinations. Amplification of genomic DNA of the 30 genotypes including 4 checks was done. Ten polymorphic RAPD primers were used to study the DNA polymorphism in some elite blackgram genotypes. All the chosen primers amplified fragments across the 30 genotypes studied, with the number of amplified fragments ranging from 5 (LC 97) to 12 (LC 73 and LC 110) and which varied in size from 200 bp to 2500 bp. Of the 89 amplifies bands 68 were polymorphic, with an average of 6.8 polymorphic fragments/ primer. Percentage polymorphism ranged from 60.0 per cent (LC-97) to a maximum of 83.3 per cent (LC-73), with an average of 75.71 per cent polymorphism. Only three out of 10 primers showed more than 80 per cent polymorphism. The study confirned the absence of sufficient genotypic variability among elite lines, emanating from intraspecific crosses in blackgram. Most of the present day varieties and lines have undergone introgression of genes either directly from T9 (a widely adapted cultivar of past) or indirectly from selection or hybridization involving T9. Better environment alone cannot lead to better yields from inferior varieties beyond a certain limit, the limit being set by the genetic makeup of the variety. The narrow genetic base of elite lines revealed the presence of low genotypic variation which suggested the need of using wide hybridization to broaden the genetic base in blackgram. Inter-specific crosses with mungbean or with wild progenitors augmented with different tools of biotechnology may contribute significantly to achieve this goal.

Introduction

Blackgram (*Vigna mungo* (L.) Hepper.), 2n=2x=22) is one of the important pulse crops in India. The major constraints in achieving higher yield of blackgram are lack of genetic variability, absence of suitable genotypes for different cropping systems, poor harvest index and susceptibility to diseases. Therefore, there is need to develop improved blackgram cultivars utilizing available genetic diversity. For efficient use of genetic variability, in plant breeding programme, it is necessary to evaluate genetic diversity and construct linkage maps (Paterson *et al.*, 1991). But to measure genetic relationship and diversity, a large number of polymorphic markers are required with reliability but in the use of morphological characters isozymes provide limited information due to their limited number and adequate levels. Molecular genetic markers have developed into a powerful tool to analyze genetic relationship and genetic diversity. The present study was conducted to investigate molecular analysis of similarity or dissimilarity using RAPD technique.

Materials and Methods

Thirty genotypes of blackgram along with 4 checks namely Pant U-19, Pant U-35, Pant U-31 and Pant U-40 of blackgram were selected for the present analysis. The name and pedigree of test genotypes along with checks of blackgram has been given in Table 6.1.

Table 6.1: List of Genotypes along their Pedigree Used in the Study

Group	*Origin*	*No. of Genotypes*	*Genotype Name*
I	Pant U30 × UPU 97-10	8	PU04-25, PU06-1, PU06-2, PU06-3, PU06-4, PU04-6, PU06-6, PU06-7
II	Pant U19 × UPU 97-10	8	PU04-29, PU04-33, PU06-8, PU06-9, PU06-10, PU06-11, PU06-12, PU06-13
III	Pant U30 × UK-17	5	PU02-12, PU04-10, PU04-12, PU04-24, PU04-31
IV	Pant U19 × UK-17	3	PU04-27, PU04-34, PU06-5
V	UPU 97-10 × Pant U 30	1	PU04-28
VI	Pant U35 × DPU 88-31	1	PU02-3
VII	UPU 1× UPU 2	1	Pant U 19
	UPU 3×Pant U 19	1	Pant U-35
	UPU 97-10× DPU 88-31	1	Pant U-31
	UPU 89-6-7×7668/4B	1	Pant U-40
	Total	30	

15 days old leafs of blackgram were taken from each genotypes grown in glass house. The fresh leaf materials were washed with autoclaved double distilled water and then with absolute alcohol. Half-a-gram leaf material was grinded in liquid N_2 to fine powder using pre-chilled pestle and mortar and then transferred the powder to Oakridge tube containing 5 ml of pre-warmed extraction buffer using spatula to dispense the material completely. Then samples were incubated at 60°C for 30 minutes with occasional mixing by gentle stirring. Three milliliter of chloroform and isoamyl alcohol (24:1) were added to each sample and mixed by inversion to emulsify spin at 15,000 rpm at room temperature for 10 minutes. The aqueous

phase was removed with a wide bore pipette and then transferred to a clean tube. Iso-propanol was added 2/3rd volume and was mixed by quick gentle stirring and then put into a deep freezer over night for 2 hours. The supernatant was spined at 10,000 rpm for 10 minutes to precipitate DNA. The DNA was then washed in 70 per cent ethanol for 20 minutes. The pellet was dried and dissolved in 500 µl TE buffers which then was stored at -20°C. RNAase treatment was given with 2 µl of RNAase. A stock was added to the eppendof tube containing 200 µl of extracted DNA and then incubated for 3 hour at 37°C in a water bath. Agarose 0.8 was dissolved in 100ml of 0.5x TBE buffer then boiling at transparent after slight cooling. Ethidium bromide 4µl was added and then comb was fitted and the solution poured on plate and waited for some time to take out the comb and load sample to run sample.

Table 6.2: List of RAPD Primer

Sl.No.	Primer Code	Primer Name	Sequence (5' to 3')
1.	LC110	Pr 41(GBP	ggACTgCAgA
2.	LC114	Pr 45(GBP)	gACgCCACAC
3.	LC118	Pr 49(GBP)	TCAgAgCgCC
4.	LC120	Pr 51(GBP)	ACAACTgCTC
5.	LC121	Pr 52(GBP)	TgACTgACgC
6.	LC96	Pr 27(GBP)	TTggCACggg
7.	LC97	Pr 28(GBP)	gTgTgCCCCA
8.	LC72	Pr 3(GBP)	AgTCAgCCAg
9.	LC73	Pr 4(GBP)	AATCgggCTg
10.	LC74	Pr 5(GBP)	AggggTCTTg

PCR amplification (Williams *et al.*, 1990) was performed with random decamer primers obtained from Operon Technologies (Almeda, Calif., USA). Amplifications were performed in a 25 µl reaction mixture containing 10 mM Tris-HCl (pH 8.3), 50 mM KCl, 2.5mM $MgCl_2$, 0.01 per cent gelatin, 0.2mM each of dNTP, 20 pmol RAPD primer, 50ng genomic DNA and 0.5 U Taq DNA polymerase (Bangalore Genei, Bangalore, India). Amplifications were performed in an Eppendorf Master cycler gradient (Eppendorf Netheler-Hinz, Hamburg, Germany). Amplified products were separated on 1.5 per cent agarose gel in 1×TBE buffer (100 mM Tris HCl, pH 8.3, 83 mM boric acid, 1 mM EDTA) at 50 V. The gel was stained with 0.5 µg/ml ehidium bromide solution and visualized by illumination under UV light. The size of amplification products were determined by comparison to λ DNA digested with *EcoR1*and *HindIII.*

DNA profile data of RAPD marker systems have been presented by using the term 'assay unit'. Assay unit is defined as one reaction involving one primer. The PCR products from RAPD analysis was scored qualitatively for presence or absence. Only clear and apparently unambiguous bands were scored for RAPD. The RAPD bands were scored as present (1) or absent (0), each of which was treated as an independent character regardless of its intensity. Binary data were used to analyze using NTSYS-pc (Numerical Taxonomy System, Version 2.11F, Rohlf 2005). The SIMQUAL subprogramme was used to calculate the Jaccard's coefficient, a common estimator of genetic identity or similarity and was calculated as follows:

$$\text{Jaccard's coefficient} = \frac{N_{AB}}{(N_{AB} + N_A + N_B)}$$

where,

N_{AB} is the number of bands shared by samples, N_A represents amplified fragments in sample A, and N_B represents fragments in sample B. Similarity matrices based on this indices were calculated. Similarity matrices were used to construct the UPGMA (un-weighted pair group method with arithmetic average) dendrograms to elucidate the diversity among the genotype studied. Correlation between the tree and similarity matrices was estimated by means of the Mantel matrix correspondence test (Mantel, 1967). This test results in product moment correlation(r) that validate the dendrogram obtained from the similarity matrix. For matrix correlation of this type, a correlation value (r) greater than 0.5 will be statistically significant at 0.01 probability level if the number of observed taxonomic unit exceeds 15 (Lapointe and Legendre, 1992).

Results and Discussion

DNA profile data for RAPD marker systems has been presented by using the term assay unit. Amplification of genomic DNA of the 30 genotypes, using 10 primers for RAPD analysis yielded 89 fragments which were scored. All the chosen primers amplified fragments across the 30 genotypes studied, with the number of amplified fragments ranging from 5 (LC 97) to 12 (LC 73 and LC 110) and which varied in size from 200 bp to 2500 bp. Of the 89 amplified bands 68 were polymorphic, with an average of 6.8 polymorphic fragments/primer. Percentage polymorphism ranged from 60.0 per cent (LC-97) to a maximum of 83.3 per cent (LC-73), with an average of 75.71 per cent polymorphism. Only three out of 10 primers showed more than 80 per cent polymorphism. Figure 6.1 is the representative of the extent of polymorphism observed among the blackgram genotypes as revealed by LC-73 primer. While in Figures 6.1 and 6.2 (a and b) polymorphism by primers LC-71 and LC- 74 have been shown.

A dendrogram based on UPGMA analysis grouped the 30 genotyped into 4 main clusters, with Jaccard's similarity coefficient ranging from 0.63 to 1.00. Two genotypes (PU06-1 and PU04-25) grouped in two different farthest clusters- cluster I and cluster II respectively, while rest 28 genotypes were differentiated into one largest (cluster III) comprising 23 genotypes and another cluster (II) comprising 5 genotypes. The largest cluster was further subdivided into two subclusters namely, subcluster II-1 consisting of 21 genotypes and subcluster II-2 consisting 2 genotypes (PU06-5 and PU06-6) Within subcluster II-1 genotypes PU04-12, PU06-10, PU06-13, PU06-11 and PU06-12 exhibited similarity coefficient value '1' and these four genotypes exhibited equal similarity coefficients with all other genotypes. Similarly, PU06-4 and PU06-9 also showed similarity coefficient value '1'. Same value of similarity coefficient was observed between Pant U 19 and Pant U 35(0.96); between PU06-7 and PU06-8 (0.96); PU06-5 and PU06-6 (0.96); Pant U 31 and PU06-3(0.98); PU04-6 and PU04-28(0.98).

The evolution of varieties in distinct agro-climatic zones demonstrates significant levels of variation in response to the selection in the zones (Singh *et al.*, 1988). Sometimes breeder also applies different selection criteria within progenies of inter-varietal crosses during recombination breeding. It is, therefore not surprising to find significant levels of polymorphism among the 30 genotypes of blackgram in RAPD markers. The RAPD technique has been applied to assess molecular polymorphism in *Vigna* (Kaga *et al.*,1996), mungbean (Santalla *et al.*, 1998; Lakhanpaul, 2000), chickpea (Sonnante *et al.*, 1997) and blackgram (Souframanien and Gopalakrishna, 2004), pea (Simioniue *et al.*, 2002), pigeonpea

Figure 6.1: The Random Amplified Polymorphic DNA (RAPD) Profile of 30 Blackgram Genotypes Including 4 Check Varieties. Amplification was done with primer LC 73(5'-AATCGGGCTG-3'). Lane M denotes λ DNA marker double digested with *EcoRI* and *Hind III*. Different colour of the blackgram entries denotes their similiarity of origin from intra specific crosses.

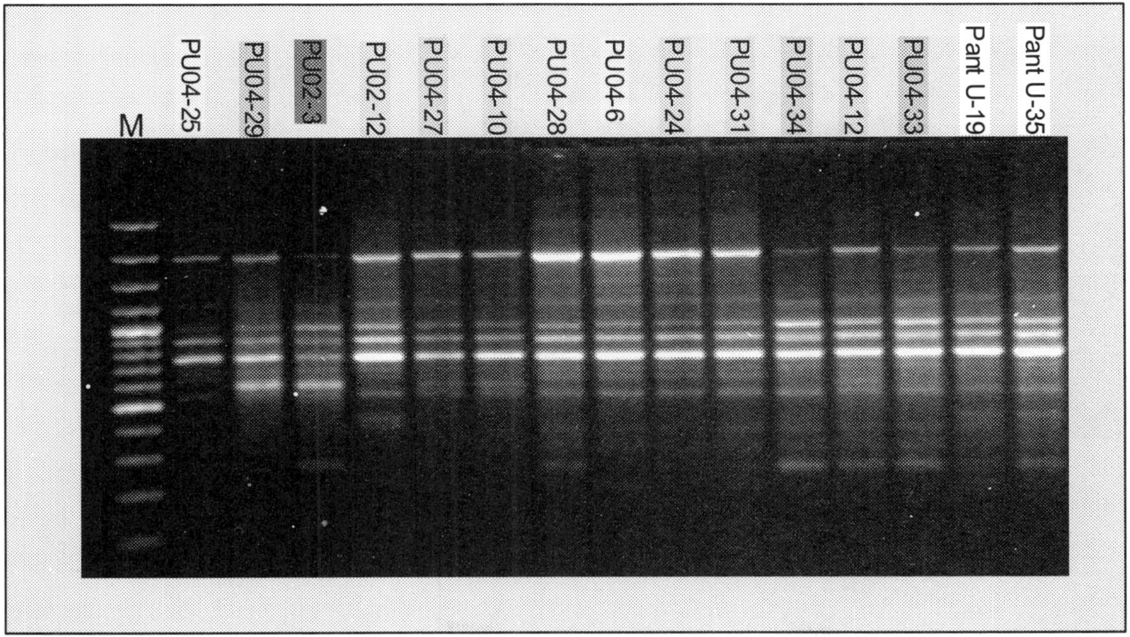

contd...

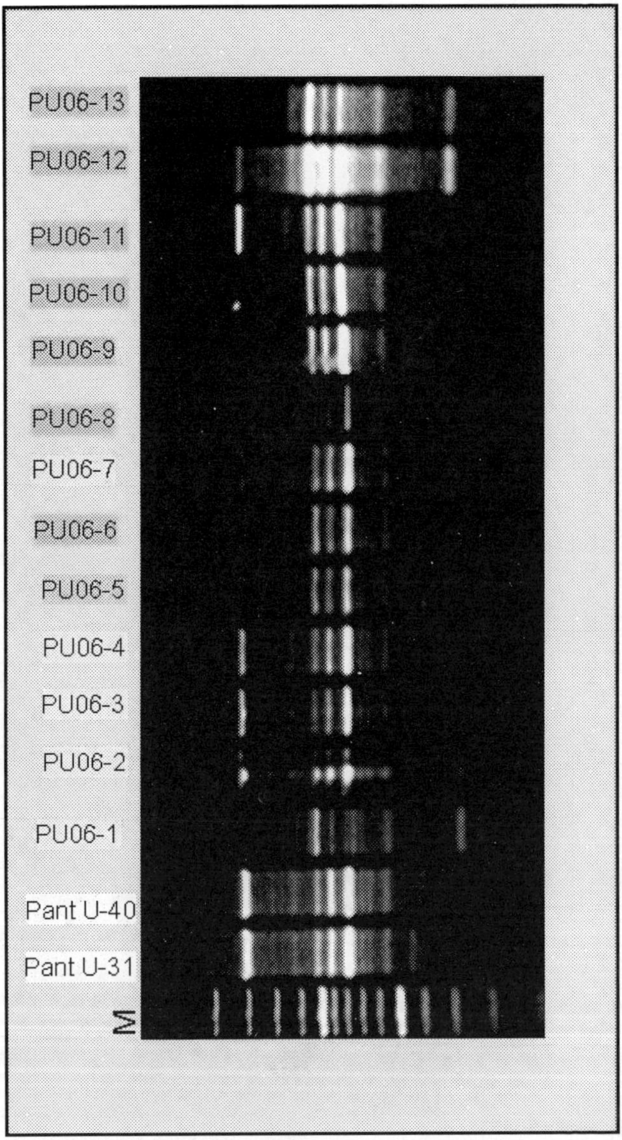

Figure 6.1—*Contd...*

Table 6.2: Representative Scoring of Bands from RAPD Profile of 30 Genotypes of Blackgram with Primer LC 73 (5'AATCGGCTTG-3')

1	PU04-25	0	1	0	0	1	1	1	1	1	1	1	0	0	0	0	1	0	0	1	1	1	1	1	1	0	0	0	1	0	0
2*	PU04-29	0	1	0	0	1	1	1	1	1	1	1	1	1	1	0	1	1	1	1	1	1	1	1	1	0	0	1	1	0	0
3	PU02-3	0	1	0	0	1	1	1	1	0	0	1	1	1	1	0	1	0	1	1	0	1	1	1	1	0	0	1	1	0	1
4	PU02-12	0	1	0	0	1	1	1	0	1	1	1	1	1	1	1	1	0	1	1	1	1	1	1	1	0	0	0	1	1	0
5	PU04-27	0	1	0	0	1	1	1	1	1	1	1	1	1	1	0	1	0	1	1	1	1	1	1	1	0	0	1	1	0	0
6	PU04-10	0	1	0	0	1	1	1	1	1	0	1	1	1	1	0	1	1	1	1	1	1	1	1	1	0	0	1	1	0	0
7	PU04-28	1	1	0	0	1	1	1	1	1	0	0	1	1	1	0	1	1	1	1	1	1	1	1	1	1	1	0	1	1	1
8	PU04-6	1	1	0	0	1	1	1	1	1	0	0	1	1	1	0	1	1	1	1	1	1	1	1	1	1	1	0	1	1	0
9	PU04-24	1	1	1	0	1	1	1	1	1	0	0	1	1	1	0	1	1	1	1	1	1	1	1	1	1	1	0	1	1	0
10	PU04-31	1	1	0	0	1	1	1	1	1	0	1	1	1	1	0	1	1	1	1	1	1	1	1	1	1	1	0	1	1	0
11	PU04-34	0	1	0	0	1	1	1	1	1	0	1	1	1	1	0	1	1	1	1	1	1	1	1	1	1	1	0	1	1	1
12	PU04-12	0	1	0	0	1	1	1	1	1	0	1	1	1	1	0	1	1	1	1	1	1	1	1	1	1	1	1	1	1	1
13	PU04-33	0	1	0	0	1	1	1	1	1	0	1	1	1	1	0	1	1	1	1	1	1	1	1	1	1	1	0	1	1	1
14	PantU-19	1	1	0	0	1	1	1	1	1	0	1	1	1	1	1	1	1	1	1	1	1	1	1	1	1	1	0	1	1	1
15	PantU-35	0	1	0	0	1	1	1	0	1	0	1	1	1	1	1	1	1	1	1	1	1	1	1	1	1	1	0	1	1	1
16	PantU-31	1	1	0	0	1	1	1	1	1	0	1	1	1	1	1	1	1	1	1	1	1	1	1	1	1	1	1	1	1	0
17	PantU-40	1	1	1	0	1	1	1	1	1	0	1	1	1	1	0	1	1	1	1	1	1	1	1	1	1	1	1	1	1	1

Contd...

Table 6.2–Contd...

18	PU06-1	0	1	0	1	0	1	1	0	0	1	1	1	1	1	0	1	1	1	0	1	1	1	1	1	1	0	1	1	1	1
19	PU06-2	0	1	0	0	1	1	1	1	1	0	0	1	1	1	0	1	1	1	1	1	1	1	1	1	1	1	1	1	1	1
20	PU06-3	1	1	0	0	1	1	1	1	1	0	1	1	1	1	0	1	1	1	1	1	1	1	1	1	1	1	1	1	1	0
21	PU06-4	0	1	0	0	1	1	1	1	1	0	1	1	1	1	0	1	1	1	1	1	1	1	1	1	1	1	1	1	1	0
22	PU06-5	0	1	0	0	1	1	1	1	1	0	1	1	1	1	0	1	1	1	0	0	1	1	1	1	1	1	1	1	1	1
23	PU06-6	0	0	0	0	1	1	1	1	1	0	1	1	1	1	1	1	1	1	0	0	1	1	1	1	1	1	1	1	1	1
24	PU06-7	0	1	0	0	1	1	1	1	1	1	1	1	1	1	1	1	1	1	1	1	1	1	1	1	1	1	1	1	1	0
25	PU06-8	0	1	0	0	1	1	1	1	1	1	1	1	1	1	0	1	1	1	1	0	1	1	1	1	1	1	1	1	1	0
26	PU06-9	0	1	0	0	1	1	1	1	1	0	1	1	1	1	0	1	1	1	1	1	1	1	1	1	1	1	1	1	1	0
27	PU06-10	0	1	0	0	1	1	1	1	1	0	1	1	1	1	0	1	1	1	1	1	1	1	1	1	1	1	1	1	1	1
28	PU06-11	0	1	0	0	1	1	1	1	1	0	1	1	1	1	0	1	1	1	1	1	1	1	1	1	1	1	1	1	1	1
29	PU06-12	0	1	0	0	1	1	1	1	1	0	1	1	1	1	0	1	1	1	1	1	1	1	1	1	1	1	1	1	1	1
30	PU06-13	0	1	0	0	1	1	1	1	1	0	1	1	1	1	0	1	1	1	1	1	1	1	1	1	1	1	1	1	1	1

Note: presence of band is denoted by '1" and absence of band by '0'

r= 0.903.

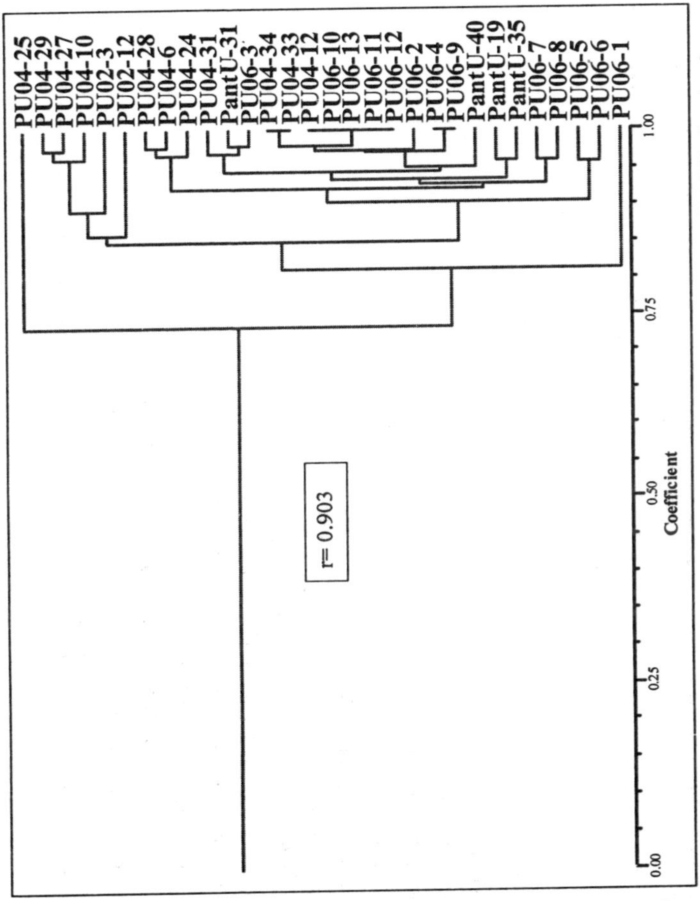

Figure 5.2: Dendrogram of Blackgram Genotypes Based on RAPD Analysis.
'r' is the cophenetic correlation coefficient of Mentel test revealing the accuracy of the tree diagram.

Table 6.3: Similarity Matrix of 30 Blackgram Genotypes Using Jaccard's Coefficients Calculated from RAPD Banding Pattern Using Ten Polymorphic Markers

Sl. No.	*Genotypes*	*1*	*2*	*3*	*4*	*5*	*6*	*7*	*8*	*9*	*10*	*11*	*12*	*13*	*14*	*15*	*16*	*17*	*18*	*19*	*20*	*21*	*22*	*23*	*24*	*25*
1.	PU04-25	1.00																								
2.	PU04-29	0.84	1.00																							
3.	PU02-3	0.74	0.88	1.00																						
4.	PU02-12	0.81	0.88	0.80	1.00																					
5.	PU04-27	0.86	0.98	0.90	0.90	1.00																				
6.	PU04-10	0.81	0.98	0.90	0.86	0.95	1.00																			
7.	PU04-28	0.70	0.83	0.79	0.80	0.80	0.84	1.00																		
8.	PU04-6	0.72	0.84	0.76	0.82	0.82	0.86	0.98	1.00																	
9.	PU04-24	0.70	0.83	0.74	0.80	0.80	0.84	0.96	0.98	1.00																
10.	PU04-31	0.75	0.87	0.79	0.84	0.84	0.89	0.96	0.98	0.96	1.00															
11.	PU04-34	0.75	0.87	0.84	0.84	0.84	0.89	0.96	0.94	0.92	0.96	1.00														
12.	PU04-12	0.73	0.89	0.86	0.83	0.87	0.91	0.94	0.92	0.90	0.94	0.98	1.00													
13.	PU04-33	0.75	0.87	0.84	0.84	0.84	0.89	0.96	0.94	0.92	0.96	1.00	0.98	1.00												
14.	PantU-19	0.71	0.83	0.80	0.85	0.81	0.85	0.96	0.94	0.92	0.96	0.96	0.94	0.96	1.00											
15.	PantU-35	0.70	0.83	0.79	0.89	0.80	0.84	0.92	0.89	0.88	0.92	0.96	0.94	0.96	0.96	1.00										
16.	PantU-31	0.71	0.88	0.80	0.85	0.85	0.89	0.92	0.94	0.92	0.96	0.92	0.94	0.92	0.96	0.92	1.00									
17.	PantU-40	0.70	0.86	0.83	0.79	0.83	0.88	0.94	0.92	0.94	0.94	0.94	0.96	0.94	0.94	0.90	0.94	1.00								

Contd...

Table 6.3–Contd...

Sl. No.	Genotypes	1	2	3	4	5	6	7	8	9	10	11	12	13	14	15	16	17	18	19	20	21	22	23	24	25
18.	PU06-1	0.63	0.82	0.78	0.79	0.79	0.79	0.78	0.76	0.74	0.78	0.83	0.85	0.83	0.79	0.83	0.79	0.82	1.00							
19.	PU06-2	0.70	0.87	0.84	0.80	0.84	0.89	0.96	0.94	0.92	0.92	0.96	0.98	0.96	0.92	0.92	0.92	0.94	0.83	1.00						
20.	PU06-3	0.73	0.89	0.82	0.83	0.87	0.91	0.94	0.96	0.94	0.98	0.94	0.96	0.94	0.94	0.90	0.98	0.96	0.81	0.94	1.00					
21.	PU06-4	0.75	0.91	0.84	0.84	0.89	0.93	0.92	0.94	0.92	0.96	0.96	0.98	0.96	0.92	0.92	0.96	0.94	0.83	0.96	0.98	1.00				
22.	PU06-5	0.67	0.84	0.86	0.77	0.82	0.86	0.89	0.87	0.85	0.89	0.94	0.96	0.94	0.90	0.89	0.90	0.92	0.84	0.94	0.92	0.94	1.00			
23.	PU06-6	0.62	0.80	0.81	0.77	0.77	0.82	0.85	0.83	0.81	0.85	0.89	0.92	0.89	0.90	0.89	0.90	0.88	0.80	0.89	0.88	0.89	0.96	1.00		
24.	PU06-7	0.76	0.92	0.80	0.89	0.89	0.89	0.88	0.90	0.88	0.92	0.92	0.94	0.92	0.92	0.92	0.96	0.91	0.83	0.92	0.94	0.96	0.90	0.90	1.00	
25.	PU06-8	0.75	0.91	0.84	0.84	0.89	0.89	0.88	0.89	0.88	0.92	0.92	0.94	0.92	0.88	0.88	0.92	0.90	0.83	0.92	0.94	0.96	0.94	0.89	0.96	1.00
26.	PU06-9	0.75	0.91	0.84	0.84	0.89	0.93	0.92	0.94	0.92	0.96	0.96	0.98	0.96	0.92	0.92	0.96	0.94	0.83	0.96	0.98	1.00	0.94	0.89	0.96	0.96
27.	PU06-10	0.73	0.89	0.86	0.83	0.87	0.91	0.94	0.92	0.90	0.94	0.98	1.00	0.98	0.94	0.94	0.94	0.96	0.85	0.98	0.96	0.98	0.96	0.92	0.94	0.94
28.	PU06-11	0.73	0.89	0.86	0.83	0.87	0.91	0.94	0.92	0.90	0.94	0.98	1.00	0.98	0.94	0.94	0.94	0.96	0.85	0.98	0.96	0.98	0.96	0.92	0.94	0.94
29.	PU06-12	0.73	0.89	0.86	0.83	0.87	0.91	0.94	0.92	0.90	0.94	0.98	1.00	0.98	0.94	0.94	0.94	0.96	0.85	0.98	0.96	0.98	0.96	0.92	0.94	0.94
30.	PU06-13	0.73	0.89	0.86	0.83	0.87	0.91	0.94	0.92	0.90	0.94	0.98	1.00	0.98	0.94	0.94	0.94	0.96	0.85	0.98	0.96	0.98	0.96	0.92	0.94	0.94

26.	PU06-9	26	27	28	29	30
27.	PU06-10	1.00				
28.	PU06-11	0.98	1.00			
29.	PU06-12	0.98	1.00	1.00		
30.	PU06-13	0.98	1.00	1.00	1.00	
26.	PU06-9	0.98	1.00	1.00	1.00	1.00

(Ratnaparkhe *et al.*, 1995), and cowpea (Mignouna *et al.*, 1998). The success of this study in identifying polymorphism is due to the use of a number of randomly selected prescreened highly informative primers.

The 10 RAPD primers in the present study yielded 89 polymorphic markers that unambiguously discriminated 30 genotypes into 4 clusters. It is evident from the RAPD derived dendrogram that except PU04-25, all other genotypes were similar to the extent of more than 75 per cent. Similarly genotype PU06-1 separated from others at a 81 per cent similarity. Surprisingly, rest 28 genotypes were differentiated into one largest cluster (cluster III) comprising 23 genotypes and another cluster (II) comprising 5 genotypes. This indicated that the variation observed in morphological characters is not being translated by molecular variation. It may also be noted that molecular variation as revealed by RAPD technique may not always be applicable because it does not reflect the functional relationship of genome with character expression of the traits in plant.

The RAPD dendrogram did not exhibit discriminating clusters according to the origin of different intra-specific crosses (groups). All elite genotypes and checks exhibited more than 75 per cent similarity and in most of the genotypes similarity ranges from 90-100 per cent. The higher degree of commonness in this pedigree may explain the high degree of closeness among genotypes studied. This is confirmed by 100 per cent similarity among two checks Pant U-19 and Pant U-35 being progenies of the cross UPU 1 ×UPU2 and UPU3×Pant U 19 respectively. Similar results were obtained in blackgram by Gaffor *et al.* (2001) and Souframanien and Gopalkrishna (2004).

Geographically isolated populations accumulate genetic differences as they adapt to different environment. Genetic variation among elite genotypes of blackgram based on RAPD analysis could be useful to select parents to be crossed for generating appropriate populations intended for both genome mapping and breeding programme.

Clustering of genotypes within group was not similar when RAPD derived dendrogram was compared with pedigree derived grouping. This difference may be attributed to marker sampling error and/or the level of polymorphism detected, reinforcing again the importance of the number of loci and their coverage of overall genome in obtaining reliable estimates of genetic relationship among cultivars (Loarce *et al.*, 1996). The putatively similar bands originating for RAPDs in different individuals are not necessarily homologous, although they may share the same size in base pair. This situation misleads us while calculating genetic relationships (Fernandez *et al.*, 2001).

Dendrograms did not indicate any clear pattern of clustering according to the pedigree of intervarietal cross from which they have been derived. Similar results have been obtained in Adzuki bean (Yee *et al.*, 1999) and groundnut (Dwivedi *et al.*, 2001).

RAPD analysis also confirmed the presence of non-significant variation among groups. Based on grain yield, maturity duration and other characters five genotypes namely; PU 04-25, PU 06-2, PU 06-4 (from group I), PU 04-10 (group III) and PU 04-28 (group V) were identified, for further testing. The narrow genetic base of the lines studied revealed by low variations between groups; this suggesting the need of incorporation of hidden genes from wild ancestors in order to broaden the genetic base in blackgram. Inter-specific crosses with mungbean or with wild progenitors may contribute significantly to achieve this goal.

DNA profile data for RAPD marker systems has been presented by using the term assay unit. Amplification of genomic DNA of the 30 genotypes, using 10 primers for RAPD analysis yielded 89 fragments which were scored. All the chosen primers amplified fragments across the 30 genotypes studied, with the number of amplified fragments ranging from 5 (LC 97) to 12 (LC 73 and LC 110) and which varied in size from 200 bp to 2500 bp. Of the 89 amplifies bands 68 were polymorphic, with an average of 6.8 polymorphic fragments/primer. Percentage polymorphism ranged from 60.0 per cent (LC-97) to a maximum of 83.3 per cent (LC-73) with an average of 75.71 per cent polymorphism.

Only three out of 10 primers showed more than 80 per cent polymorphism. Two genotypes (PU06-1 and PU04-25) grouped in two different farthest clusters- cluster I and cluster II, respectively. While rest 28 genotypes were differentiated into one largest (cluster III) comprising 23 genotypes and another cluster (II) comprising 5 genotypes. The largest cluster was further subdivided into two subclusters namely, subcluster II-1 consisting of 21 genotypes and subcluster II-2 consisting 2 genotypes (PU06-5 and PU06-6). Within subcluster II-1, genotypes PU04-12, PU06-10, PU06-13, PU06-11 and PU06-12 exhibited similarity coefficient value '1' and these four genotypes exhibited equal similarity coefficients with all other genotypes. Similarly, PU06-4 and PU06-9 also showed similarity coefficient value '1'. Same value of similarity coefficient was observed between Pant U-19 and Pant U-35 (0.96); between PU06-7 and PU06-8 (0.96); PU06-5 and PU06-6 (0.96); Pant U-31 and PU06-3 (0.98); PU04-6 and PU04-28(0.98). The high amount of similarity among the genotypes is the result of derivation of some genotypes from common ancestors. Even the different parents contributing genes into elites lines were somehow related either closely or distantly to a widely adaptable blackgram variety of the past *i.e.* Type 9.

References

Dwivedi, S. L., Gurtu, S., Chandra, S., Yuejin, W. and Nigam, N. 2001. Assessment of genetic diversity among selected groundnut germplasm: rapd analysis. *Plant Breeding*, 120: 345-349.

Fernandez, M. E., Figueiras, A. M. and Benitoc. 2002. The use of ISSR and RAPD markers for detecting DNA polymorphism, genotype-identification and genetic diversity among barley cultivars with known origin. *Theoretical and Applied Genetics*, 104: 845-851.

Ghafoor, A., Sharif, A., Ahmad, Z., Zahid, M. A. and Rabbani, M.A. 2001. Genetic diversity in blackgram (*Vigna mungo* l. (hepper). *Field Crops Research*, 69(2): 183-190.

Kaga, A., Tomooka, N.; Egawa, Y.; Hosaka, K. and Kamijima. 1996. Species relationships in the subgenus *ceratotropis* (*genus Vigna*) as revealed by rapd analysis. *Euphytica*, 88(7): 17-24.

Lakhanpaul, S., Chandhu, S. and Bhat, K.V. 2000. Random amplified polymorphic dna (rapd) analysis in indian mung bean (*Vigna radiata*). *Genetica*, 109(8): 227-234.

Loarce, Y., Gallego, R. and Ferrer, E. 1996. A comparative analysis of the genetic relationships between rye cultivars using rflp and rapd markers. *Euphytica*, 88: 107-115.

Lopointe, F.J. and Legendre, P. 1992. Statistical significance of the matrix correlation coefficient for comparing independent phylogenetic trees. *Syst. Biol*, 41: 378-384.

Mignouna, H.D., Ikca, J. and Thottapilly, G. 1998. Genetic diversity in cowpea as revealed by random amplified polymorphic DNA. *Genet. Breed.*, 52: 151-159.

Paterson, Hirshi, A. and Claudio, S. 1991. Purine and pyrimidine metabolism in cultured white spruce (picea glauca) cells: metabolic fate of C^{14} labeled precursors and activity of key enzymes. *Physiologia Plantarum*, 108(1): 25-33.

Ratnaparkhe. 1995. A comparative analysis of genetic diversity in blackgram genotypes using rapd and issr markers. *Theoretical and Applied Genetics*, 109(8): 176-184.

Rohlf, F.J. 2005. Ntsys-pc numerical taxonomy and multivariate analysis system version 2.02i. *Exter Publication*, 44(2): 76-79.

Santalla, M., Power, J.B. and Davey, M.R. 1998. Genetic diversity in mungbean germplasm revealed by rapd markers. *Plant Breeding*, 117(5): 473-478.

Simioniue, S., Saini A., Reddy, K.S. and Jawall, N. 2002. Evalution of long primers for ap-pcr analysis of mungbean (*Vigna radiata* (L.) Wilczek): genetic relationships and fingerprinting of some genotypes. *Indian Journal Biotechnology*, 3: 511-518.

Singh, I. 1998. Variability and correlation in blackgram. *Crop Research*, 16(2): 220-223.

Sonnante G., Piergiovanni, A. R. and Perrino. 1996. Relationships of *Vigna unguiculata, v. Vexillata* and species of section *vigna* based on isozyme variation. *Genetic Resources and Crop Evolution*, 43: 157-165.

Souframanien, J. and Gopalakrishna, T. 2004. A comparative analysis of genetic diversity in blackgram genotypes using rapd and issr markers. *Theoretical and Applied Genetics*, 109(8): 1687-1693.

Williams, J. G.K., Kubelik, A.R., Livak, K.J., Rafalski, J. A. and Tingsey, S.V. 1990. DNA polymorphisms amplified by arbitrary primers are useful as genetic markers. *Nucleic Acids Research*, 18: 6531-6535.

Yee, E., Kidwell, K.K., Sills, G.R. and Lumpkin, T.A. 1999. Diversity among selected *Vigna Angularis* accessions on the basis of rapd and aflp markers. *Crop Sciences*, 39: 268-275.

Chapter 7

Physiological Characterizations of Nitrogen Use Efficiency in Different Genotypes of Rice (*Oryza sativa* L.) to Reducing Environmental Pollution

☆ *Hukum Singh, Amit Verma, Bhawana Joshi, Pawanika Chandola and Alok Shukla*

Excessive N fertilization creates environmental hazards like nitrate pollution in surface water bodies. Therefore, an optimum application of N fertilizer is required in order to protect the crop from N deficiency and also to curtail environmental pollution. A field experiment was conducted to evaluate the physiological characterizations of Nitrogen Use Efficiency in different genotypes of rice on alluvial soil of Dr. N.E. Borlaug Crop Research Centre and Department of Plant Physiology, G.B. Pant University of Agriculture and Technology Pantnagar, Uttarakhand, India. This experiment was performed during rainy season of 2008 and 2009. Five rice genotypes, *viz.*, Vasumati (aromatic), Tulsi, Kasturi, Krishna Hamsa (high yielding) and KRH-2 (hybrid) were used as experimental material. The fertilizer urea was taken as nitrogenous source and 0, 50, 100 and 200N kg/ha of nitrogen fertilizer was sprouted in the field. The evaluated physiological parameters of different rice genotypes showed the positive correlation between different nitrogen levels. Leaf and seed nitrogen content, NR activity, photosynthetic rate were increased with increasing nitrogen levels. The maximum N content in leaf was found to be for KRH-2 (1.783 per cent) hybrid rice genotype while minimum for Vasumati (1.460 per cent) aromatic rice genotype. The nitrogen content in seed was highest in Krishna Hamsa (1.452 per cent) and lowest for KRH-2 (1.285 per cent). Photosynthetic rate was maximum in Vasumati (8.11 μmol $m^{-2} s^{-1}$) but minimum for Tulsi (7.34 μmol $m^{-2} s^{-1}$). The rice genotype Krishna Hamsa showed the better response to nitrogen fertilizer for better rice yield. In future, these studies will be of help in optimizing doses of Nitrogen fertilizer in cultivation of rice crop and pollution can be controlled.

Introduction

Nitrogen is universally the most warning nutrient for crop production in the bulk of world's agricultural areas. Efficient use of nitrogen is also vital for minimizing environmental pollution. Excessive use of nitrogen fertilizers results in decrease of physiological nitrogen-use efficiency and causes serious environmental pollution (Jiang *et al.*, 2005). At low N accessibility, C_3 plants have a better PNUE and NUE than C_4 plants, whereas at high N, the opposite is true (Sage *et al.*, 1987). Nutrient use efficiency is the ability to utilize the available Nutrient resources to optimize productivity of the crops. In agriculture, 60 per cent Nitrogen is used for cultivation of cereal crops in which only 30 per cent N is utilized by crops and remaining 70 per cent losses through volatilization, ammonification, leaching and some denitrifying bacteria and, as a result, causes so many environmental problems like acidification, global warming and eutrophication. Efficient utilization of nitrogen fertilizers is essential for ensuring better yield as well as for minimizing the adverse impacts of nitrogen species in the environment (Cao *et al.*, 1984; Balasubramanian *et al.*, 1998).

The physiological impact of plant N accumulation with respect to an increased photosynthetic activity requires critical consideration as a supplemental investment of N in the photosynthetic machinery and may be detrimental to the transfer of N to the grain and thus to final yield (Sinclair *et al.*, 2004). At high leaf N contents, the maximum rate of individual leaf photosynthesis per unit leaf area in C_3 plants can be as high as that in C_4 plants with lower N contents. So, leaves of individual C_3 plants can have high rates of photosynthesis comparable to those of C_4 leaves (Sage *et al.*, 1987).

Reduction in N concentration per leaf area increases the light-use efficiency but decreases the overall net photosynthesis. It was also argued that photosynthesis might be the least affected process by N deficiency in agricultural production long back by Wong (1979). Higher nitrogen content and uptake in plant organs due to supra-optimal N application or soil extraction can reduce the utilization efficiency for maximum grain production (Swain *et al.*, 2006).

Leaf photosynthesis and leaf nitrogen content are closely related. A decline in the photosynthetic rate has been reported as a consequence of both lower leaf nitrogen concentration and some other factors such as the amount and activity of Rubisco protein. Greater nitrogen supply results in elevated nitrogen and Rubisco per unit leaf area in developing leaves (Mae, 1997). The nitrogen content of leaf is an important contribution to the photosynthetic rate and the photosynthetic nitrogen-use efficiency. Three factors that are important for photosynthetic function are leaf chlorophyll content, chlorophyll a:b ratio and total leaf nitrogen content (TLN) (Hu *et al.*, 2006).

Nitrate Reductase is a complex cytosolic enzyme made up of two identical subunits catalyzing the transfer of two electrons from NAD(P)H to NO_3^- ion via several redox centers composed of three prosthetic groups, *viz.* flavin adenine dinucleotide (FAD), heme (cytochrome 557) and a molybdenum pterin cofactor (Mo-Co). Nitrate Reductase activity in rice varies with seeding age, genotype, nitrogen-form and concentration of nitrogen supply. Leaf nitrate reductase activity was found to be maximum at the time of maximum tillering and lowest during flowering (Reddy *et al.*, 1988). The activity of nitrate reductase is a measure of the activity of higher plants to take up, and use nitrate (Ouko *et al.*, 2003). A positive correlation was observed in nitrate reductase activity and grain protein and in leaf protein of rice (Sharma and Ruchi, 2005).

Thus, the goal of this study was to evaluate photosynthetic rate, NR activity and nitrogen content of leaves and seeds in five rice genotypes at different nitrogen levels because these parameters are directly associated with climate changes and global warming. Further these findings will be helpful in future research for selection of high nitrogen efficient rice genotype and using these genotypes, environment problems can be tackled, and energy and a lot of money can be saved.

Materials and Methods

The present study was performed during rainy season 2008 and 2009 at the Dr. N. E. Borlaug Crop Research Centre and Department of Plant Physiology, G. B. Pant University of Agriculture and Technology Pantnagar, Uttarakhand using the five rice genotypes as the experimental material. The objective of the study was to assess the effect of various nitrogen levels (N_0, N_{50}, N_{100} and N_{200} kg/ha) on Photosynthetic rate, NR activity and N content from the leaves and grains after harvesting. The seeds of five rice genotypes, namely, KRH-2 (hybrid), Vasumati (aromatic), Kasturi, Tulsi, Krishna Hamsa (high yielding varieties) were obtained from Directorate of Rice Research, Rajendranagar, Hyderabad. The urea was used as a nitrogen source. The experimental field was laid out in split plot design with three replications.

Nitrate Reductase (NR) activity was estimated in vivo in freshly harvested flag leaves at active tillering and flowering by using the method described by Hageman and Hucklesby (1971). Photosynthetic rate was measured at flowering by Infra-red Gas Analyzer (IRGA) (TPS-2, PP system, USA). Nitrogen content of rice grains and leaves was estimated as per procedure of Association of Official Agricultural Chemists (1970).

Result and Discussion

Nitrogen Content (Per cent)

The effect of different N levels on N (per cent) was estimated in leaves at the time of flowering stage and grain after harvesting and it's presented in Table 7.1. The percentage of nitrogen content increased with increasing doses of fertilizer nitrogen. The maximum N (per cent) at N_0 level in leaf was found to be for the rice genotype KRH-2 (1.40) and the minimum for rice genotype Kasturi (1.22). At

Table 7.1: Effect of Different N Levels on Nitrogen Content (Per cent) in Leaf at Flowering and Grains After Harvesting in Rice Genotypes

Genotype	Nitrogen (Per cent)									
	Leaf at Flowering					Grain After Harvesting				
	N_0	N_{50}	N_{100}	N_{200}	Mean	N_0	N_{50}	N_{100}	N_{200}	Mean
KRH-2	1.40	1.75	1.85	2.15	1.783	1.09	1.09	1.38	1.58	1.285
Vasumati	1.35	1.40	1.46	1.65	1.460	1.10	1.35	1.56	1.57	1.394
Kasturi	1.22	1.70	1.63	1.72	1.563	1.20	1.48	1.47	1.58	1.430
Tulsi	1.28	1.50	1.52	1.71	1.498	1.22	1.31	1.58	1.66	1.443
Krishna Hamsa	1.24	1.55	1.74	1.75	1.565	1.22	1.34	1.53	1.71	1.452
Mean	1.293	1.575	1.635	1.791		1.164	1.315	1.504	1.620	
		Treatment (T)	Variety (V)		T × V		Treatment (T)	Variety (V)		T × V
S.Em.±		0.0006	0.0002		0.0006		0.0038	0.0047		0.0095
CD at 5 per cent		0.0022	0.0008		0.0016		0.0133	0.0137		0.0275

N_{50} level the maximum N (per cent) in leaf was found to be for rice genotype KRH-2 (1.75) and the minimum for rice genotype Vasumati (1.40). At N_{100} level the maximum N (per cent) in leaf was found to be for rice genotype KRH-2 (1.85) and the minimum for rice genotype Vasumati (1.46). The maximum leaf N (per cent) at N_{200} level was reported in rice genotype KRH-2 (2.15) and the minimum in rice genotype Vasumati (1.5). The overall genotypic mean recorded highest N (per cent) in leaf for rice genotype KRH-2 (1.783) and the lowest for rice genotype Vasumati (1.460). Under greater N amount, rice plants distribute more N in leaf blades (Jiang *et al.*, 2005). The leaf, stem and grains nitrogen content by the medium and late duration rice varieties increased with increasing N rates and for all the varieties the stem N content was the lowest followed by leaf and grain. It was found to be 5.2 gm/kg in stem, 6.4 gm/kg in leaf and 11.2gm/kg in grains (Swain *et al.*, 2006).

The maximum N (per cent) at N_0 level in grain was found to be for the variety rice genotype Krishna Hamsa (1.22) and the minimum for rice genotype KRH-2 (1.09). At N_{50} level the maximum N (per cent) in grains was found to be for rice genotype Kasturi (1.48) and the minimum for rice genotype KRH-2 (1.09). At N_{100} level the maximum N (per cent) in grains was found to be for rice genotype Tulsi (1.58) and the minimum for rice genotype KRH-2 (1.38). The maximum N (per cent) at N_{200} level was reported in rice genotype Krishna Hamsa (1.71) and the minimum in rice genotype Vasumati (1.57). The overall genotypic mean recorded highest N (per cent) in grains for rice genotype Tulsi (1.443) and the lowest for rice genotype KRH-2 (1.285). Similarly, it was reported by Stalin *et al.* (1999) that the N content in grains and straw increased with increasing N levels. In grains it was found 0.79 per cent and in straw 0.64 per cent at N_0 and 1 per cent in grains and 0.79 per cent in straw at N_{200} level. The increase in N per cent with increasing N levels and from straw to grain was also reported to be 1.15 per cent in grains and 0.38 per cent in straw at N_0 treatment while it was 1.19 per cent in grain and 0.50 per cent in straw at N_{200} level (Roul and Sarawgi, 2005).

The different forms of N fertilizers had significant effect on grain N. A plant treated with sulfur containing N fertilizers had the more grain N than non-sulfur containing N fertilizers. This might be due to the role of sulfur in protein synthesis (Chaturvedi, 2005).

Nitrate Reductase Enzyme Activity (µmol NO_2-g^{-1} fresh wt. h^{-1})

The effect of different nitrogen levels on leaf nitrate reductase activity at active tillering and flowering is presented in Table 7.2. The NR activity increased with increasing nitrogen levels but decreased towards increasing plant age. It was found to be high during active tillering stage than during flowering stage. During active tillering stage the maximum NR activity at N_0 level was found to be for rice genotype Tulsi (0.103) and the minimum for rice genotype Kasturi (0.067). At N_{50} level it was found to be highest for rice genotype Vasumati (0.135) and the minimum for rice genotype Tulsi (0.075). At N_{100} level the maximum NR activity was found to be for rice genotype Vasumati (0.185) and the minimum for rice genotype Tulsi (0.083).

The maximum NR activity at N_{200} level was found to be for rice genotype Krishna Hamsa (0.307) and the minimum for rice genotype Tulsi (0.148). The overall genotypic mean showed that the Vasumati (0.164) had the maximum NR activity while the variety rice genotype Tulsi (0.102) had the minimum at active tillering. T x V interaction was found to be statistically significant for rice genotype KRH-2 between N_{50}-N_{200} and for rice genotype Tulsi between N_{100}-N_{200} while for rice genotypes Vasumati, Kasturi and Krishna Hamsa at N_0-N_{200} N levels.

During flowering stage the NR activity at N_0 level was recorded higher for rice genotype Tulsi (0.078) and the lower for rice genotype KRH-2 (0.050). At N_{50} level the NR activity was found to be greatest for rice genotype Vasumati (0.113) and rice genotype Kasturi (0.113) and the lowest for rice genotype Tulsi (0.074).

The utmost NR activity at N_{100} level was found to be for rice genotype Vasumati (0.154) and the least for rice genotype Tulsi (0.077). At N_{200} level the maximum NR activity was observed in rice genotype KRH-2 (0.359) and the minimum in rice genotype Tulsi (0.123). The overall genotypic pattern showed that the rice genotype KRH-2 (0.151) had the maximum NR activity and the rice genotype Tulsi (0.088) had the minimum at flowering stage. Similarly Reddy *et al.* (1988) reported that nitrate reductase activity was maximum at active tillering (150 n mol NO_2^- h^{-1}) followed by panicle initiation and flowering (110 n mol $NO_2^-h^{-1}$ and 90 n mol $NO_2^-h^{-1}$) respectively.

Table 7.2: Effect of Different N Levels on Nitrate Reductase Activity (µmol NO_2^- g^{-1} fresh $wt.h^{-1}$) at Active Tillering and Flowering Stage of Five Rice Genotypes

Genotype	*Nitrate Reductase Activity*									
	At Active Tillering					*At Flowering*				
	N_0	N_{50}	N_{100}	N_{200}	*Mean*	N_0	N_{50}	N_{100}	N_{200}	*Mean*
KRH-2	0.051	0.087	0.180	0.275	0.148	0.050	0.093	0.102	0.359	0.151
Vasumati	0.088	0.135	0.185	0.248	0.164	0.063	0.113	0.154	0.169	0.125
Kasturi	0.067	0.105	0.115	0.212	0.125	0.074	0.113	0.101	0.132	0.105
Tulsi	0.103	0.075	0.083	0.148	0.102	0.078	0.074	0.077	0.123	0.088
Krishna Hamsa	0.069	0.112	0.118	0.307	0.152	0.057	0.109	0.104	0.160	0.108
Mean	0.076	0.103	0.136	0.238		0.064	0.100	0.108	0.189	
	Treatment (T)		*Variety (V)*		*T × V*	*Treatment (T)*		*Variety (V)*		*T × V*
S.Em.±	0.0043		0.0063		0.0126	0.0183		0.0207		0.0415
CD at 5 per cent	0.0149		0.0182		0.0364	0.0634		0.0597		0.1195

The plants with more N uptake and more nitrate reductase activity have higher concentration of protein (Sharma and Ruchi, 2005).

Table 7.3: Effect of Different N Levels on Photosynthesis Rate ($\mu mol\ m^{-2}\ s^{-1}$) at Flowering Stage of Five Rice Genotypes

Genotype	*Photosynthesis Rate at Flowering ($\mu mol\ m^{-2}\ s^{-1}$)*				
	N_0	N_{50}	N_{100}	N_{200}	*Mean*
KRH-2	6.98	7.67	7.66	8.05	7.59
Vasumati	7.70	7.36	7.89	9.48	8.11
Kasturi	7.23	7.27	7.15	8.20	7.46
Tulsi	6.93	6.96	7.47	8.00	7.34
Krishna Hamsa	6.62	7.57	7.30	8.05	7.39
Mean	7.09	7.37	7.49	8.36	

	Treatment (T)	*Variety (V)*	*T × V*
S.Em.±	0.081	0.107	0.215
CD at 5 per cent	0.282	0.310	0.621

Photosynthetic Rate at Flowering ($\mu mol\ m^{-2}\ s^{-1}$)

Table 7.3 shows the photosynthetic rate at different N levels at flowering stage. All the genotypes of rice showed the high rate of photosynthesis at flowering. The maximum rate of photosynthesis at N_0 level was found to be for rice genotype Vasumati (7.70) and the minimum for rice genotype Krishna Hamsa (6.62). At N_{50} level the maximum photosynthetic rate was recorded for rice genotype KRH-2 (7.67) and the minimum for rice genotype Tulsi (6.96). The maximum photosynthetic rate at N_{100} level was found to be for rice genotype Vasumati (7.89) and the minimum for Kasturi (7.15). At N_{200} level the maximum photosynthetic rate was observed in rice genotype Vasumati (9.84) and the minimum for rice genotype Tulsi (8.00). The overall genotypic mean showed the maximum photosynthetic rate for variety rice genotype Vasumati (8.11) and the minimum for rice genotype Tulsi (7.34).

The increased photosynthetic rate might be due to the fact that N is an integral component of photosynthetic machinery, so it was increased with increasing N levels from N_0 to N_{200}. It is also reported that the split application of N from N_1 to N_4 increases the dry matter accumulation in two rice cultivars. This increased dry matter accumulation which could be due to increased photosynthetic rate with increasing N levels from N_1 to N_4. Many studies have well documented photosynthetic responses of plants to N availability, which included significant and positive correlation between photosynthetic capacity and leaf N content, suggesting that most of the nitrogen is used for the synthesis of components in the photosynthetic apparatus (Sugiharto *et al.*, 1990; Egli and Schmid 1999; Shangguan *et al.*, 2000; DaMatta *et al.*, 2002; Makoto and Koike, 2007).

References

Association of Official Agricultural Chemists Washington. 1970. *Official Method of Analysis of the Official Agricultural Chemists*. Edn. 11. pp. 934. Association of Official Agricultural Chemists, Washington DC.

Balasubramanian, V. Morales, A.C., Cruz, R.T. and Abdul Rachman, S. 1998. On-farm adaptation of knowledge-intensive nitrogen management technologies for rice systems. 1-12. In: *Nutrient Recycling in Agroecosystems*. Kluwer Acad. Publ. Netherlands.

Cao, Z.H., De Datta, S.K., and Fillery, I.R.P. 1984. Nitrogen-15 balance and residual balance of urea-N in wetland rice fields as affected by deep placement techniques. *American Journal of Soil Science*. 48: 203-208.

Chaturvedi, I. 2005. Effect of nitrogenous fertilizers on growth, yield and quality of hybrid rice. *Journal of Central Europian Agriculture*. 4: 611-612.

DaMatta, F.M., Loos, R.A., Silva, E.A., Loureiro, M.E. and Ducatti, C. 2002. Effects of soil water deficit and nitrogen nutrition on water relations and photosynthesis of pot-grown *Coffea canephora* Pierre. *Trees*.16: 555-558.

Egli, P. and Schmid, B. 1999. Relationships between leaf nitrogen and limitations of photosynthesis in canopies of *Solidago altissima*. *Acta oecol*. 20: 559-570,

Hagman, R.H. and Hucklesby, D.P. 1971. Nitrate reductase from higher plants. *Methods of Enzymology*. 23: 491-503.

Hu, M.L., Wang, C.M., Yang, Q.H., Zhai, H.Q. and Wan, J.M. 2006. QTL analysis for traits associated with photosynthetic function in rice (*Oryza sativa* L.). *Research Notes III. Genetics of Physiological Traits and Others.* 21. 205-220.

Jiang, L., Dong, D., Gan, X. and Wei, S. 2005. Photosynthetic efficiency and nitrogen distribution under different nitrogen management and relationship with physiological nitrogen use efficiency in three rice genotypes. *Plant and Soil.* 271:321-328.

Mae, T. 1997. Physiological nitrogen efficiency in rice: nitrogen utilization, photosynthesis, and yield potential. In: Ando T, ed. *Plant Nutrition for Sustainable Food Production and Environment.* Dordrecht, The Netherlands: Kluwer Academic Publishers, pp. 51-60.

Makoto, K. and Koike, T. 2007. Effects of nitrogen supply on photosynthetic and anatomical changes in current-year needles of *Pinus koraiensis* seedlings grown under two irradiances. *Photosynthetica.* 45: 99-104.

Ouko, M., Asch, F. and Becker, M. 2003. Nitrate reductase activity in rice as related to weed competitiveness. http://www.arts.uni-bonn.de.

Reddy, A.T., Reddy, D.V.R., Rao, V.S., Kumar, T.V., Pillai, R.N. and Rao, I.V.S. 1988. Effect of nitrogen forms on leaf nitrate reductase activity, yield and protein content of rice. *International Rice Research Notes.* 13: 23-24.

Roul, P.K. and Sarawgi, S.K. 2005. Effect of integrated nutrient techniques on yield, N_2 content, uptake and use efficiency of rice (*Oryza sativa*). *Indian Journal of Agronomy.* 50:129-131.

Sage, R.F., Pearcy, R.W. and Seeman, J.R. 1987. The nitrogen use efficiency in C_3 and C_4 plants. *Plant Physiology.* 85:355-359.

Shangguan, Z.P., Shao, M.A. and Dyckmans, J. 2000. Nitrogen nutrition and water stress effects on leaf photosynthetic gas exchange and water use efficiency in winter wheat. *Environ. Experimental Botany.* 44: 141-149.

Sharma, N. and Ruchi. 2005. Nitrogen assimilation potential, periodic nitrate reductase activity and its relationship to grain proteins in field grown rice in presence of butachlor. *Indian Journal of Plant Physiology.* 10: 196-198.

Sinclair, T.R., Purcell, L.C. and Sneller, C.H. 2004. Crop transformation and the challenge to increase yield potential. *Trends in Plant Science*. 9:70-75.

Stalin, P., Thiyagarajan, T.M. and Ragarajan, R. 1999. Nitrogen application strategy and use efficiency in rice. *Oryza*. 36:322-326.

Sugiharto, B., Miyata, K., Nakamoto, H., Sasakawa, H. and Sugiyama, T. 1990. Regulation of expression of carbon-assimilating enzymes by nitrogen in maize leaf. *Plant Physiology*. 92: 963-969.

Swain, D.K., Bhaskar, B.C., Krishnan, P., Rao, K.S., Nayak, S.K. and Dash, R.N. 2006. Variation in yield, N uptake and N use efficiency of medium and late duration rice varieties. *Journal of Agriculture Science*. 144: 69-83.

Wong, S.C. 1979. Elevated atmospheric partial pressure of CO_2 and plant growth. I. Interactions of nitrogen nutrition and photosynthetic capacity in C_3 and C_4 plants. *Oecologia*. 44: 68-74.

Chapter 8

Elucidation of Photothermic Behaviour of Different Rice (*Oryza sativa* L.) Varieties for Yield

☆ *Jyoti, Alok Shukla, Deepti Shankhdhar and S.C. Shankhdhar*

Rice is the staple food for more than half of the world's population. The world population is about 5.8 billion and is going to touch 7.5 billion by 2020. To fulfill the need of population rice production must be increased by 50 percent to compensate its growing demand. The majority of rice in India is grown during the wet season (July - November) in the traditional rice growing areas. During this season the major limiting factors for higher productivity are sunlight (duration and intensity) and temperature. In rice production, the light and temperature combinations play an important role since its light intensity requirement is higher and temperature-dependent. A field experiment was performed to evaluate the photothermic behavior of six rice genotypes, namely IET 21075,

IET 21105, IET 21119, Tulasi, Govind, and Annada in two planting conditions, *i.e.* normal (early) and after 15 days (late) sown. Various parameters like plant height, tiller/plant, total dry matter (TDM), primary and secondary branches/panicle, 1000 grain weight, harvest index, and chlorophyll content were recorded in early and late sown conditions. All the six genotypes showed different photothermic behavior for yield and their yield attributes. Amongst the genotypes, IET 21075 and IET 21119 showed maximum TDM. The maximum days to panicle initiation were also recorded in IET 21075 and IET 21105 in early and late sown conditions respectively. IET 21105 showed maximum number of primary and secondary branches in early sown conditions as compared to late sown conditions. The panicle weight, economic yield and harvest index were also recorded maximum for IET 21105 and IET 21075 in early and late sown conditions respectively. 50.85 per cent and 27.96 per cent increase in yield during early sown as compared to late sown conditions were recorded in IET21105 and Govind respectively. The sensitivity of rice genotypes for photoperiod in terms of yield was as IET21105> Annada> IET21119>I ET21075> Tulasi>Govind. Overall all the varieties performed better in early sown conditions.

Introduction

Rice is an important and primary source of food for more than one third of world's population. India ranks first in terms of area under rice production and second largest country in terms of production with an average yield of 2.8 tons per hectare. To fulfill the demand of growing population rice production must be increased. Annual changes in weather can cause yield variation of nearly 30 per cent. Breeding new elite cultivars and using the best management techniques, 18 per cent increase in yield is estimated in the coming decades by avoiding the negative effect of climate change (Sheehy *et al.*, 2005). There are many biotic factors such as pathogens, weeds and disease and abiotic factors such as light, temperature, time of sowing which affect the production of rice (Reddy, 2006). In rice production, the light and temperature combinations play an important role since its light intensity requirement is higher and temperature dependent (Venkataraman, 1987).

Rice is sensitive to temperature fluctuation but its sensitivity depends on developmental stages to extreme temperature with very cold and very hot weather leading to high spikelet sterility. Also, high air temperature reduces tillering, causes stunting and accelerates the developmental rate of rice thereby leading to shortened growing cycles (Futakuchi, 2005). Flowering (anthesis and fertilization) and to an extent booting (microsporogenesis), are the most susceptible stages of development to temperature in rice (Farrell *et al.*, 2006). Sterility is caused by poor anther dehiscence and low pollen production, and hence low numbers of germinating pollen grains on the stigma (Matsui *et al.*, 2001). The rice productivity is adversely affected due to elevation in mean minimum night temperature as it reduces crop duration, increases respiration rates and evapotranspiration, alters photosynthate partitioning to grains, decreases fertilizer use efficiency and accelerates nutrient mineralization in soil (Pathak and Ladha, 2006). The low temperature is one of the major constrains in rice (*Oryza sativa* L.) production that reduces the grain yield all over the world. In different geographical regions rice varieties often get chilling injury of various degrees and types, especially in the south China, which is double-season rice area in China. The most sensitive period of rice to low temperature is at the seedling and flowering stages. As a consequence of increased concentration of CO_2 and greenhouse gases, earth's temperature is increasing. With an increase in temperature above normal, the phenological development of crops is advanced and yield is decreased. An increase of 1.0°C in temperature leads to 10 per cent decrease in yield. In addition, decrease in solar radiation causes decreased yield and vice-versa (Hundal and Kaur, 2007). The present study was taken up to find out the suitability of rice genotypes under climatic changes, for such suitability will help fulfill the food grain needs of the growing population.

Materials and Methods

The seeds of six rice genotypes, namely, IET21075, IET21105, IET21119, TULASI, GOVIND, ANNADA were used and obtained from the Directorate of Rice Research, Hyderabad, India. The seeds of rice were sown in two different dates *i.e.*, normal (early) sowing and 15days delayed sowing thereafter 21 days old seedlings of normal and late sowing conditions were then transplanted to the prepared field.

Statistical Design and Layout

The field experiment was carried out with two treatments and three replications of each in a split plot design. The plots were separated from each other with proper spacing (20 x 20) and the experimental field was bordered by proper bunds. The field was in flowed by one main and one sub- channel.

Fertilizer Dose

100 N – 45 P_2O_5-60K_2O Kg/ha. (P&K as a basal dose)

N- Splits

1st 50 per cent of scheduled N at 10- 15 days after planting, 2nd 25 per cent of scheduled N at active tillering, 3rd 25 per cent of scheduled N at panicle initiation (PI)

Agronomical/Morphological and Physiological Characters

Three competitive plants were randomly selected from each replication. Observations were recorded on per plot basis. Height of the plant was measured in cm. from soil base up to the base of topmost fully expanded leaf. Total number of green leaves was recorded at the time of tillering. Days to panicle initiation were recorded by calculating the number of days from the date of sowing to the date of appearance of panicle primordia. Days to general flowering were recorded by calculating the number of days from date of sowing to the date of heading. It is the stage when 50 per cent of the panicles in the plot are exerted.

Crop Growth Rate $\overline{(CGR)}$ (g m^{-2} day^{-1})

It represents total dry matter productivity of the crop community per unit land area over a certain time span.

$$\overline{(CGR)} = \frac{\text{Dry weight at flowering } (W_2) - \text{Dry weight at tillering } (W_1)}{\text{Days to flowering } (t_2) \text{ Days to tillering } (t_1) \times \text{area}}$$

Total Dry Matter

The total dry matter was calculated at maximum tillering and flowering stages by uprooting the complete plant and then placing the plant samples in the oven at 80°Cfor three days.

Yield

Each plant was uprooted from the ground level at maturity and after drying, the weight of intact plant was determined before thrashing and the total weight of the plant was recorded as biological yield in t/ha. Grain yield (economic yield) from each replication was recorded in terms of t/ha after harvesting. One thousand well developed dried grains were randomly selected from each replication and weighed. The harvest index was calculated using the following formula:

$$\text{Harvest index (per cent)} = \frac{\text{Economic yield}}{\text{Biological yield}} \times 100$$

Results and Discussion

Date of sowing is an important event in regulating the phenological traits of rice and yield attributes. In the present investigation photothermal behavior of six rice varieties, namely, IET 21075, IET 21105, IET 21119, Tulasi, Govind and Annada were observed with respect to yield and micronutrient such as iron and zinc content by changing the date of sowing *i.e.* early and late sown condition.

Plant Height

The plant height was non significantly decreased in late sown condition. Rice variety Govind recorded the maximum height (61.60 cm) under early sown condition, whereas Annada attained minimum (52.85 cm). Rice genotypey IET 21075 showed the maximum height (57.00cm) and IET 21105 (52.93cm) showed minimum height in late sown condition. All the varieties responded well during early sown condition (Table 8.1). Increase in plant height in early sown condition might be due to the availability of prolonged period for vegetative growth. With the delay in planting time, reduction in plant height was observed. Similarly in an early sown condition, rice varieties performed better than late sown condition (Halappa *et al.*, 1974). In the another study it was also observed that different genotypes of rice were affected significantly when assessed through the interaction of varieties and transplanting date (Safdar, 2008).

Table 8.1: Plant Height, Total Dry Matter/Hill at Maximum Tillering and Flowering Stage and ($\overline{CGR}$) of Different Rice Varieties During Early and Late Sown Conditions (±indicate SE)

Variety	*Plant Height (cm)*		*Total Dry Matter/Hill (g) at Tillering*		*Total Dry Matter (g) at Flowering*		*($\overline{CGR}$) (g m^{-2} day^{-1})*	
	Early	*Late*	*Early*	*Late*	*Early*	*Late*	*Early*	*Late*
IET 21075	57.92±2.91	57.0±2.51	15.95±0.77	14.39±1.02	31.74±1.32	21.44±2.30	31.16±1.31	21.11±2.31
IET21105	56.40±1.87	52.93±1.21	15.60±1.03	14.91±0.80	38.10±0.77	27.47±1.83	37.58±0.80	27.13±1.82
IET21119	57.20±4.97	55.93±5.61	15.50±1.04	16.71±1.41	44.11±2.46	21.80±1.57	43.53±2.47	21.16±1.52
Tulasi	56.93±4.62	53.60±3.22	13.23±0.86	17.93±1.21	40.10±2.04	18.55±2.41	39.28±1.98	17.81±2.44
Govind	61.60±5.87	54.86±1.81	14.49±1.07	16.82±1.04	29.10±1.35	16.66±0.34	28.20±1.38	5.93±0.37
Annada	52.85±2.91	54.86±1.81	12.82±1.40	17.48±0.95	30.88±4.02	19.00±1.73	30.13±3.94	18.27±1.76
S.Em.±	0.71	1.23	0.25	0.44	0.54	0.94	0.48	0.84
CD at 5%	2.09	3.6	0.76	1.31	1.60	2.77	1.43	2.48

Total Dry Matter

The total dry matter at tillering non-significantly differed in an early and late sown condition. The maximum dry matter (15.95 g) was observed in variety IET 21075 and the minimum was in Annada (12.82g) in an early sown condition. In the late sown condition the maximum was recorded in Tulasi (17.93g) and the minimum in IET 21075 (14.39g). The total dry matter increased in late sown at maximum tillering. The total dry matter at flowering was significantly increased in early sown condition as compared to late sown condition. In early sown condition the maximum dry matter was recorded in IET 21119 (44.11g) and minimum in Govind while in late sown condition the maximum dry matter was recorded in IET 21105 (27.47g) and minimum in Govind (Table 2.1). This might be due to variation in the light intensity in early and late sown condition, under the low light intensity the reduction in dry matter content is related to impaired photosynthesis. Dry matter partitioning has also been found to be affected by different factors such as temperature, light, nitrogen, fertilizer and irrigation supplies (Rahman *et al.*, 2007; Rauf and Sadaqat, 2008).

During growth and development of rice plants, dry mass increases along the growth cycle, and nutrient uptake is subject to great variations (Murayama, 1995). Similar results were reported by Hassan *et al.* (2003). Higher number of productive tillers in an early transplanted plots than the late transplanted and total dry matter depended on number of tiller and leaf area index. In the present study the crop growth rate was significantly high in an early sown condition as compared to late sown condition. The maximum crop growth rate was recorded in IET 21119 (43.53 $gm^{-2}d^{-1}$) and minimum in Govind (28.20$gm^{-2}d^{-1}$) in an early sown condition. In late sown condition IET 21105 (27.13 $gm^{-2}d^{-1}$) showed the maximum crop growth rate and Govind (5.93 $gm^{-2}d^{-1}$) showed the minimum (Table 2.1). This might be due to the temperature because rice $\overline{CGR}$ was highest at intermediate temperature. The regression model indicated that the crop growth rate ($\overline{CGR}$) was adversely affected by the ambient temperature and photosynthetic active radiation (PAR) within the crop canopy.

Days to Panicle initiation and Days to 50 per cent Flowering

The panicle initiation days were significantly different in an early and late sown condition. The maximum days to panicle were recorded in IET 21075 (70.66 days) and the minimum in Annada (55.66 days) in an early sown condition and in late sown condition IET 21105 (67.00 days) recorded the maximum days to panicle initiation (Table 8.2). Photoperiod influenced the crop during restricted plant stage while temperature affected plant growth at all stages (Roberts and Summerfield, 1987). Photoperiod control of flowering showed differential response between species to species. Continuous exposure of long days in rice cultivar delays panicle emergence. Early transplanted crops received more favorable temperature especially at the tillering stage and panicle initiation stages compared to later transplanted crops (Singh *et al.*, 2005).

In the present study flowering days were non significantly differed in early and late sown condition. The flowering days were decreased in late sown condition. In an early sown condition Tulasi took the maximum (106 days) and in late sown condition varieties IET 21075 took the maximum (92.33 days) to general flowering. Variety IET 21119 took the minimum (87.66 days) in an early sown condition and Govind (81.33 days) in late sown condition. This might be due to flowering of rice varieties which appears to be associated with a response to more than one environmental factor, namely average length of day and the day degrees. Similar results were reported by Tanaka (1962). Flowering usually related to daily maximum air temperatures and the optimum daily maximum air temperature is 31~32°C. When the daily maximum air temperature is below about 25°C, flowering is seriously curbed. Combination of either higher summation of degree days and shorter days or by a lower summation of degree days leads to flowering (Bose *et al.*, 1996).

Number of Primary and Secondary Branches/Panicle

Number of primary branches significantly differed in an early and late sown condition. The number of primary branches was significantly higher in variety IET 21105 (10.77) in an early sown condition while variety IET 21075 (9.40) showed significant increase in late sown condition. The minimum number of primary branches/ panicle was in IET 21119 (7.88) in an early sown condition while in late sown condition Govind (7.40) showed the minimum number of secondary branches/panicle (Table 8.2).

Table 8.2: Days to Panicle Initiation and General Flowering, Number of Primary Branches and Secondary Branches/Panicle of Different Rice Varieties during Normal (Early) and Late Sown Conditions (±indicate SE)

Varieties	*Days to Panicle Initiation*		*Days to 50% Flowering*		*Number of Primary Branches/Panicle*		*Number of Secondary Branches/Panicle*	
	Early	*Late*	*Early*	*Late*	*Early*	*Late*	*Early*	*Late*
IET 21075	70.66±0.58	57.33±5.77	88.33±.58	92.33±1.15	9.62±0.03	9.40±0.72	54.55±0.96	42.20±5.41
IET21105	67.33±1.15	67.00±1.00	90.00±0.70	90.66±1.53	10.77±0.20	8.20±0.72	31.10±4.35	27.53±2.34
IET21119	67.00±5.00	59.00±1.00	86.66±0.58	83.66±1.15	7.88±0.20	7.30±0.30	26.21±3.50	25.73±2.73
Tulasi	60.66±1.53	50.33±1.15	106.33±0.58	82.0±1.73	8.41±1.00	6.50±0.17	25.21±3.10	18.66±4.73
Govind	56.33±1.15	50.33±0.58	96.33±17.62	81.33±1.15	9.09±1.01	6.40±0.35	27.99±1.53	17.20±0.35
Annada	55.66±0.58	56.00±1.73	96.66±17.90	81.66±1.15	9.99±0.34	8.50±0.17	26.66±2.73	19.85±4.19
SEm.±	1.06	0.61	1.62	2.82	0.12	0.21	0.76	1.33
CD at 5%	3.12	1.80	4.77	8.27	0.37	0.64	2.25	3.90

The maximum number of secondary branches was significantly higher in variety IET 21075 in both early (54.55) and late (42.20) sown condition. Variety Tulasi showed the minimum (25.21) in an early sown condition and in late sown condition Govind showed minimum (17.20) number of secondary branches/panicle (Table 8.2). The primary and secondary branches were significantly high in an early sown condition as compared to the late sown condition. This might be due to the dates of transplanting, early planting enable the plant to have longer time for growth before flowering as compared to late. Similar result was reported by Trivedy and Kwatra (1983), that early transplanting genotypes showed the maximum number of primary and secondary branches.

Yield Component

1000 Grains Weight (g)

Thousand grains weight significantly varied in an early and late sown condition. It ranged from 17.0 to 25.66 g in an early and 17.33 to 24.66 g in late sown conditions. All the genotypes showed reduction in thousand grain weight in late sown as compared to normal (early) sown condition except IET 21075. IET 21075 showed slight increased in thousand grain weight, however, its overall yield did not increase as compared to normal sown condition. The thousand grains weight was maximum in Annada (25.66 and 24.33 g) and minimum in IET20175 (17.0 and 17.33 g) in both early and late sown condition respectively. Early transplanting was significantly superior for grain yield over late transplanting. This might be due to the better root density in early sown condition which favoured the uptake of nutrients and thereby increase in grain yield. As the planting was delayed, the yield components and grain yield decreased significantly (Table 8.3). Lowered grains weight in late sown condition, might be because of increased day night temperature during ripening period (Chen and Lin, 1979). Thousand-grain weight, an important yield-determining component, is a genetic character least influenced by environment (Ashraf *et al.*, 1999). Thousand grains weight was found to be significantly increased in an early condition. Similar types of results were also observed by earlier workers (Gill *et al.*, 2006). On the contrary the high panicle weight was also observed in early sown condition by Vange and Obi (2006). Panicle weight and grain yield declined significantly with increasing day/night temperature (Ziska and Manalo, 1996).

Table 8.3: 1000 Grains Weight, Economic Yield, Biological Yield and Harvest Index (Per cent) in different Rice Varieties during Early and Late Sown Conditions (±indicate SE)

Varieties	*1000 Grains Weight (g)*		*Economic Yield (t/ha)*		*Biological Yield (t/ha)*		*Harvest Index (per cent)*	
	Early	*Late*	*Early*	*Late*	*Early*	*Late*	*Early*	*Late*
IET 21075	16.00±1.00	17.33±1.15	6.69±1.6	4.48±0.19	15.66±2.52	12.9±0.29	48.48±6.67	49.59±9.13
IET21105	23.00±1.00	19.00±0.40	7.63±0.97	3.75±0.20	14.66±0.58	12.33±00	57.64±12.4	43.22±10.35
IET21119	24.66±1.15	23.00±1.00	7.46±1.26	4.40±0.42	18.66±3.06	14.83±00	43.63±5.82	45.66±6.09
Tulasi	22.66±0.58	21.33±0.58	5.28±1.06	3.76±0.45	15.00±1.0	12.16±1.15	40.49±3.98	44.14±9.0
Govind	23.33±1.53	19.66±0.58	5.58±0.77	4.02±2.27	12.33±2.08	10.83±0.76	54.25±14.9	47.03±20.1
Annada	25.66±0.89	24.66±1.53	6.53±0.88	3.85±0.97	16.00±2.65	12.75±2.50	48.39±8.02	42.52±3.71
SEm.±	0.22	0.39	0.22	0.39	0.40	0.70	2.03	2.03
CD at 5%	0.66	1.15	0.67	1.16	1.20	2.08	5.95	10.32

Spikelet sterility was increased by late transplanting due to low temperature at panicle emergence stage (Nahar *et al.*, 2009).

Fully Filled Grains (per cent), Half Filled Grains (per cent) and Chaffi Grains (per cent)

The fully filled grains were significantly different in an early and late sown condition. The maximum (per cent) of fully filled grains was observed in Tulasi (70..66 per cent) and the minimum in the IET 21075 (29.0 per cent) in an early sown condition while in late sown condition the maximum fully filled grains were found in varieties IET 21105 (63.66 per cent) and the minimum in IET 21119 (22.33 per cent) (Figure 8.1).

The partially filled grains significantly differed in an early and late sown condition. In the early sown condition Govind showed the maximum (57.66 per cent) and Tulasi showed the minimum (24.0 per cent) while IET 21119 showed the maximum (67.66 per cent) and Annada recorded the minimum (27.33 per cent) partially filled grains in late sown condition.

The chaffy grains were significantly reduced in early sown condition. In an early sown condition the minimum chaffy grains were observed in Tulasi (5.66 per cent) while in late sown condition Tulasi observed the minimum (10.0 per cent) (Figure 8.1). The grains filling was affected by dates of transplanting. The maximum fully filled grains were observed in an early sown condition. This might be due to the low temperature at panicle initiation in late sown condition that cause spikelet sterility. The similar result was observed by Munnendra, 1988. Synchronization of grain filling period with lower temperature causes grain sterility leading to decreased filled grain percentage (Janardhan *et al.*, 1980). Higher temperature leads to faster grain growth rate with shorter grain filling period resulting in reduced duration of grain filling period thus reducing the grain weight (Yoshida, 1981).

Economic Yield (t/ha), Biological Yield (t/ha) and Harvest Index (per cent)

Economic yield was significantly increased in all the varieties in early sown condition whereas in late sown condition the yield is dramatically reduced in all the six varieties of the present study. The reduction in grain yield varied from 28 per cent in Govind to 50.85 per cent in IET21105 during late sown as compared to early sown

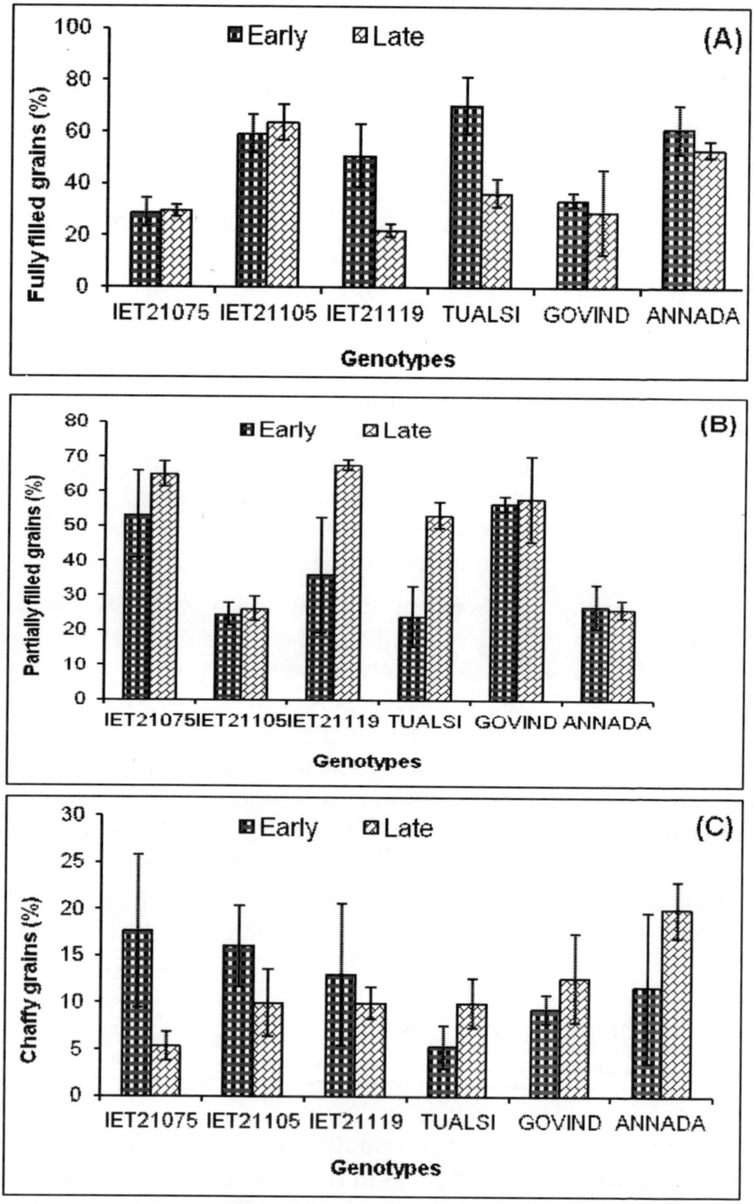

Figure 8.1: Fully Filled Grains (A), Partially Filled Grains (B) and Chaffy Grains (C) in Different Rice Varieties during Early and Pate Sown Conditions (vertical bars indicate±SE)

condition (Figure 8.2). This might be due to the favourable climatic condition especially at the time of tillering, flowering and grains filling during normal (early sown) condition. Rice grain yield decreased as the date of seedling was delayed. Similar results were observed by (Chopra *et al.*, 1998). It might be due to long sunshine hours during early sown condition, as they are important for proper grain filling condition. If sunshine hours are reduced there is ultimately decrease in the grain yield in rice (Yoshida, 1981). Low temperature during microspore development increased spikelet sterility and reduced grain yield in rice (*Oryza sativa* L.) (Farrell *et al.*, 2006).

In present study biological yield was significantly increased in an early sown condition as compared to the late sown condition. In an early sown condition variety IET 21119 showed the maximum biological yield (18.66 t/ha) and the minimum (12.33 t/ha) in Govind, while in late sown condition the maximum in IET 21119 (14.83 t/ha) and Govind showed the minimum (10.83 t/ha). The reduction in biological yield varied from 12.17 per cent in Govind to 20.53 per cent in IET 21119 in late sown as compared to normal (early sown) condition. Increase in biological yield in early sown condition might be due to exposure of plant for more sunshine hours. Straw is the product of growth during vegetative phase. Longer growth duration leads to greater biomass production and consequently greater grain yield (Ying *et al.*, 1998). Crop photosynthesis and respiration losses are the major determinants of total crop biomass, which are sensitive to temperature (Peng *et al.*, 2004).

The harvest index non significantly varied in an early and late sown condition. In both the conditions the harvest index increased non significantly. Similarly significant increase in height and spikelet sterility, reduced biological and economic yields and lower harvest index were observed by Hassan *et al.* (2003). Harvest index is also affected by date of transplanting. Increased harvest index leads to increased grain yield. Low light stress caused the reduction in biological and economic yield due to fewer grains per panicle and lower harvest index in rice (Singh, 2005). The results of present study clearly indicate that if sowing of rice plants is delayed by 15 days the biological yield is not much more affected as compared to economical yield i.e grain yield. It means reduction in grain yield

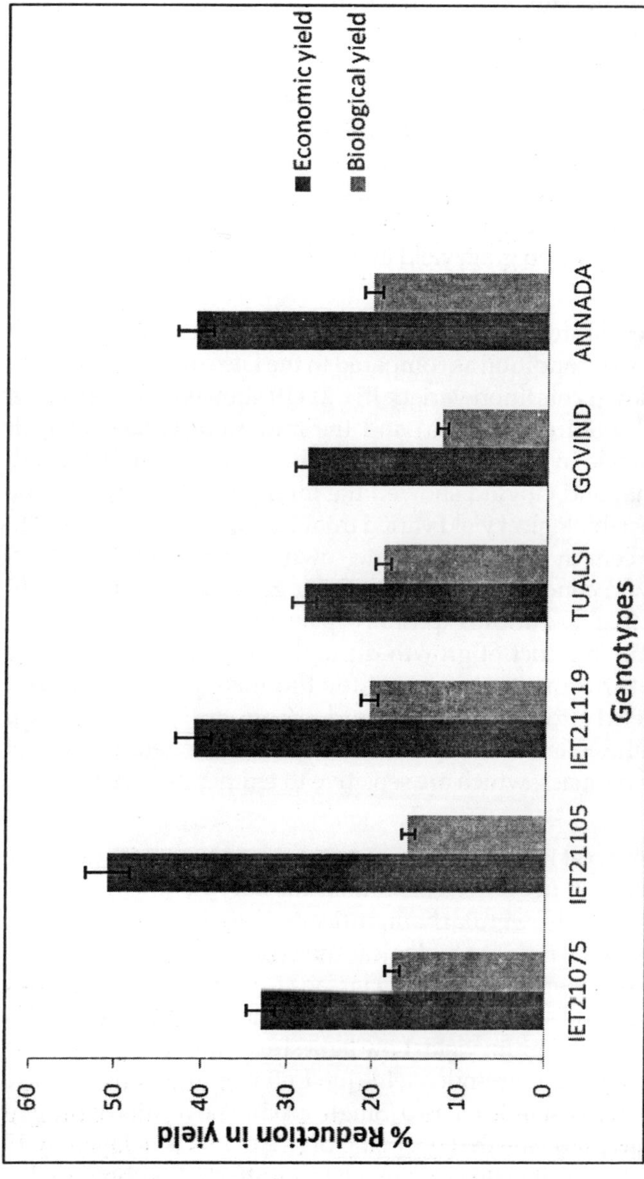

Figure 8.2: Per cent Reduction in Economic and Biological Yields in Different Rice Genotypes Grown Under Late-Sown as Compared to Normal (Early) Sown Conditions (vertical bars indicate±SE)

was due to poor translocation of photosynthates from shoots to grains and also because of sterility due to change in climatic condition during reproductive phase. The maximum panicle weight, economic yield and harvest index was recorded maximum for IET 21105 in an early sown condition while in IET 21075 in late sown. Overall all the varieties performed better in normal (early sown) condition. Some genotypes are more sensitive such as IET 21105 which showed 50.85 per cent reduction in grain yield during late sown condition while some genotypes (Govind & Tulasi) are less sensitive (28 per cent reduction in yield) to photothermic behavior. On the basis of such study Govind and Tulsi might be recommended to cultivate under climate change condition.

References

Abou-Khalif and Basyoni, A.A. 2009. Evaluation of some hybrid rice varieties under different sowing times. *African J. of Plant Sci.* 3: 53-58.

Ashraf, A., Khalid, A. and Ali, K. 1999. Effect of seedling age and density on growth and yield of rice in saline soil. *Pak. J. Biol. Sci.* 2:860-862.

Bose, M., Saha, A., Adak, M.K. and Das Gupta, D.K. 1996. Studies on relative photosynsitivity of elite dwarf indica rice cultures. *Indian J. Plant Physiol.* 1: 195-197.

Breusegem, F.V., Vranova, E., Dat, J.F. and Inze, D. 2001. The role of active oxygen species in plant signal Transduction. *Plant Sci.* 161: 405-414.

Chen, C.J. and Lin, S.S. 1979. Effect of day and night temperature on leaf appearance rate and grain weight of rice in china. *J. Agr. Assoc.* China, 95: 24-35.

Chopra, R.S., Khan, G.M., Bali, A.S. and Shah, M.H. 1998. Response of direct seeded rice to moisture regimes and sowing dates in Kashmir valley. *Oryza,* 35: 279-281.

Farrell, T.C., Fox, K.M., William, R.L., Fukai, S. and Lewin, L.G. 2006. How to improve reproductive cold tolerance of rice in Australia. *Australian J. Agri. Res.* 57: 89-100.

Futakuchi, K. 2005. Submergence damage in rice and challenges in expanding the crop's adaptability to submerged conditions in

West and central Africa. Proceeding of the World Rice Research Conference held in Tokio and Tsukuba, Japan, pp 4-7.

Gill, M.S., Kumar, A. and Kumar, P., 2006. Growth and yield of rice cultivars under various methods and time of sowing. *Indian J. Agron.* 51: 123-127.

Habibullah, N.H., Ahmad, N. and Iqbal, F. 2007. Response of rice varieties to various transplanting dates. *Pak. J. Plant Sci.* 13: 1-4.

Halappa, G., Khan, T.A., Mahadevappa, M. and Venkatasamu M.N. 1974. Optimum time of planting high yielding varieties of paddy in kharif season of tank fed tracts of Karnataka Mysore. *J. Agri. Sci.* 8: 488-492.

Hassan, G., Khan, N.U. and Khan, Q.N. 2003. Effect of transplanting date on the yield and yield components of different rice cultivars under high temperature. *Sci. Khy.* 16:129-137.

Hundal, S.S. and Kaur, P. 2007. Climatic variability and its impact on cereal productivity in Indian Punjab. *Curr. Sci.* 92: 506-511.

Janardhan, K.V., Murthy, K.S. and Das, N.D. 1980. Effect of low light during ripening period on grain yield and translocation of assimilates in rice varieties. *Indian J. Plant Physiol.* 23: 163-168.

Jiang, C.D., Gao. H., Zou, Q. and Jiang, G. 2006. Leaf orientation, Photorespiration and xanthophyll cycle protect young soybean leaves against high irradiance in field. *Environ. Exp. Bot.* 55: 87-96.

Matsui, T., Omasa, K. and Horie, T. 2001. The difference in sterility due to high temperatures during the flowering period among japonica-rice varieties. *Plant Prod. Sci.* 4: 90–93.

Muneendra, A. 1988. Effect of planting time and variety on growth and yield of rice. *Oryza.* 25: 319-322.

Murayama, N. 1995. Development and senescence. In: Matsuo, T., Kumazawa, K., Ishii, R., Ishihara, K. and Hirata, H. Science of the rice plant. *Tokyo: Physiology, Food and Agricultural Policy Research Center.* 2: p.119-132.

Nahar, K., Hasanuzzaman, M. and Majumder, R.R. 2009. Effect of Low Temperature Stress in Transplanted Aman Rice Varieties Mediated by Different Transplanting Dates. *Acad. Jour. of Plant Sci.* 2: 132-138.

Nonami, H., Wu, Y. and Boyer, J.S., 1997. Decreased growth-induced water potential (a primary cause of growth inhibition at low water potentials). *Pl. Physiol.* 114: 501-509.

Pathak, H. and Ladha, J.K. 2006. Rice: environmental issues. *Indian Farming*. pp 46-49.

Peng, S., Huang, J., Sheehy, J.E., Laza, R.C., Visperas, R.M., Zhong, X., Khush, G.S. and Cassman, K. G. 2004. Rice yields decline with higher night temperature from global warming. *Proc Natl. Acad Sci.* 101: 9971–9975.

Rahman, H.R., Malik, S.A., Saleem, M. and Hussain, F. 2007. Evaluation of seed physical traits in relation to heat tolerance in upland cotton. *Pak. J. Bot.* 39: 475-483.

Rauf, S. and Sadaqat, H.A. 2008. Effect of osmotic adjustment on root length and dry matter partitioning in sunflower (*Helianthus annuus* L.) under drought stress. *Acta Agric. Scand. Section B. Soil and Plant Sci.* 58: 252-260.

Reddy, C.K. 2006. Photo-Nyctoperiodism and Photothermics. Modern physiological data acquition, management and analysis. Short term training course. Directorate of Rice Research Rajendranagar Hyderabad.

Roberts, E.H. and Summerfield R.J. 1987. Measurement and prediction of flowering in annual crops. In J.G. Atherton (ed.) *Manipulation of Flowering*. Butterworths. London. P. 17-50.

Safdar, M.E., Ali, A., Muhammad, S., Sawar, G. and Awan, T.H. 2008. Effect of transplantins dates on paddy yield of fine grain rice genotype. *Pak. J. Bot.* 40: 2403-2411.

Sheehy, J.E., Elmido, A., Centeno, G. and Pablico, P. 2005. Searching for new plants for climate change. *J. Agric. Meteorol.* 60: 55-61.

Singh, R.P., Brennan, J.P., Farrell, T., Williams, R., Reinke, R., Lewin, L. and Mullen, L. 2005. Economic analysis of improving cold tolerance in rice in Australia. Cooperative Research Centre for Sustainable Rice Production, Australia.

Singh, V.P., Dey, S.K. and Murty, K.S. 1988. Effect of low light stress on growth and yield of rice. *Indian J. Plant Physiol.* 31: 84-91.

Tanaka, A. 1962. Nutrient absorption by rice root under submersed conditions. *J. Sci. Soil Manure.* 5: 43-48.

Trivedy, K.K. and Kwatra, K.L. 1983. Effect of dates of transplanting and hill spacing on growth and yield of rice. *JNKVV Res. J.* 17: 227-231.

Vange, T. and Obi, I.U. 2006. Effect of Planting Date on Some Agronomic Traits and Grain Yield of Upland Rice Varieties at Makurdi, Benue State, Nigeria. *J. of Sustainable Develop. in Agri. And Environ.* 2: 794-867.

Venkatraman, R. 1987. Studies on thermophotosensitivity of paddy plant under field conditions. *Indian Acad. Sci.* 59: 117-136.

Yamakawa, H., Hirose, T., Kuroda, M. and Yamaguchi, T. 2007. Comprehensive expression profiling of rice grain filling-related genes under high temperature using DNA microarray. *Plant Physiol.* 144: 258-277.

Yamazaki, J., Ohashi, A., Hashimoto, Y., Negishi, E., Kumagai, S., Kubo, T., Oikawa, T., Maruta, E. and Kamimura, Y. 2003. Effects of high light and low temperature during harsh winter on needle photodamage of *Abies mariesii* growing at the forest limit on Mt. Norikura in Central Japan. *Plant Sci.* 165: 257-264.

Ying, J., Peng, S., He, Q., Yang, H., Yang, C., Visperas, R.M. and Cassman, K.G. 1998. Comparison of high yield rice in tropical and subtropical environments I. Determinants of grain and dry matter yields. *Field Crops Res.* 57: 71-84.

Yoshida, S. 1981. Physiological analysis of rice yield. In: *Fundamentals of Rice Crop Science.* Makita City (Philippines): International Rice Research Institute, pp. 231-251.

Ziska, L. and Manalo, P. 1996. Increasing night temperature can reduce seed set and potential yield of tropical rice. *Aust. J. Plant Physiol.* 23: 791-79.

Chapter 9

Biodiversity and Gender in India

☆ *Vandana Shiva and Vir Singh*

There is an intimacy between biodiversity and women. Women take care, conserve and enhance biodiversity. They have been the managers and custodians of biodiversity. Nature itself carries a feminine notion. Women and nature, it appears, are made for each other. Women understand the value of biodiversity and do everything possible to nurture it, be it in the forests, pastures or in cultivated areas. Women's access to and control over means of production, such as land and land use, crops and animals, natural resources and ecosystems, education and training, and labour and income are limited. However, their role in production and overall socioeconomic development is significant. Food production systems where women play a leading role flourish with biodiversity, unlike the Green Revolution type agriculture operated by men which is flawed by vulnerability-ridden monocultures supported by intensive irrigation, chemical fertilizers and pesticides. Equipped with autonomy and supported by equal rights and rights to access and control over natural resources, the women's contribution to natural resources' development, food production, food sovereignty and security, biodiversity conservation and enhancement and climate change mitigation would be phenomenal.

An enormous number of names were applied by native agriculturists to soils. Each district rejoices in at least a dozen purely local names. Native cultivators keenly appreciate the smallest differences in the relative qualities of different soils. Names have references to external condition, frequency or decency of cultivation, its situation as regards inhabited sites, etc., its position as upland or recently formed alluvium, its occupation for pastures, fields or gardens, its external features, etc.

– A.O. Hume (1878)

Generation after generation local knowledge about the natural processes and resources has been transmitted. Women's role in land use has been essential in this context. This not only concerns food production, but also water and fuel supply and the provision of fodder, fibres, fruits, and other natural products. The land use methods which have been developed by these women – such as the use of natural fertilizers, seed selection, soil conservation, crop rotation, mulching, and terracing – are well-adapted towards natural circumstances. Sir Howard, the 'father' of modern sustainable agriculture, already mentioned these gender specific roles in his *Agricultural Testament* of 1940.

It is also obvious that especially where modern technology has been introduced, the perceptions of local women of the natural environment are often more multi dimensional as compared to those of men (Hoskins, 1979). Based on existing power structures and the differences in gender roles and tasks, men often prefer commercial returns instead of welfare returns for the family.

Whereas a forest acts in the eyes of a woman as a multifunctional system, men tend to focus on one or only a few outputs. Western science and technology are based on a similar reductionistic view and approach. In agriculture and forestry this has resulted in a promotion of monocultures and a loss of crop diversity. The differences in interests between men and women could also lead to a distinction in the perception of the environment by men and women. However, there might also be a great difference in the environmental perception depending on class, age, and environmental situation.

The more holistic perception, which sees different matters in a mutual relationship, has been called a feminine perception and a vision which is reductionistic, a masculine perception. The present power structures, which are dominated by (Western) men, make that a masculine perception is often related to men and a feminine vision is often coupled to women, but that is not a precondition.

An important indicator to describe the position of women and gender aspects is a level of women's autonomy. Schrijvers (1986) defines "autonomy" in this context as: "the scope that women have under varying circumstances in which to live their own lives". The introduction of technologies or other interventions could have a positive or a negative impact on the autonomy and position of women.

Lnportant criteria for the autonomy of women are:

- ✩ Her access to and control over means of production, such as land and land use, crops and animals, natural resources and ecosystems, education and training, and labour and income;
- ✩ Forms of cooperation and organisation of women which will enable and help them to control their own affairs;
- ✩ Their control over their own sexuality and fertility;
- ✩ Gender conceptions which legitimate women's sense of dignity and self-respect and their right to self-determination (Schrijvers, 1986), including woman's perception of her own situation and environment.

Although in this case much attention has to be paid to gender differentiation with regard to women's access to and control over means of production, the other criteria are also essential to determine the effects of technological introductions.

Women and Biodiversity

They (women) have a profound knowledge of the plants, animals and ecological processes around them.

–Irene Dankelman and Joan Davidson (1988)

Several studies have already shown that women in Africa, Asia and – to a somewhat lesser extent – in Latin America, play a predominant role in the use and management of natural resources; it is in agriculture, use and management of forest products (like fodder and fuel) and the supply of water. Apart from agricultural and allied field activities, she performs all other household chores. Children, especially girls, assist her.

In several areas in India the key role and position of women in managing the natural cycles in the agro-ecosystem is still evident; for example, in the hill and mountain areas of North India. We will describe the situation there briefly.

Singh (1987) showed that a pair of bullocks works for 1064 hours, a man for 1212 hours and a woman for 3485 hours in a year on one hectare farm in the Indian Central Himalayas. Women in the hill agriculture in Himachal Pradesh did 37 percent of work in sowing, 59 percent in interculture, 66 percent in harvesting, 59 percent in trenching, and 69 percent in tending the animals. And all this is apart from all the household chores, which include the collection of fuel and water.

Her contributions in the animal husbandry are as follows: she harvests the crops and stacks the hay for domestic animals, she transports the leaf fodder and bedding material over long distances on difficult terrain, she grazes the cattle on distant grazing lands, carries animals to water sources for water, takes care of young calves, milks the animals, cleans the animal shed and executes all other activities related to animal husbandry, except the ploughing, castration, purchase, and sale of animals (Singh, 1987; Singh, 1998).

But also in other agricultural activities her role is evident. Women know everything about the cultivation practices of indigenous varieties of food crops. For example, in Henwal Valley of Garhwal Himalayas, where they do almost all the farm work themselves in local paddy fields, including the seed selection of indigenous rice varieties.

Gay, in a study of 1983, recorded that women rice growers in central Liberia used and recognised well over hundred varieties of rice. During an experiment, the women identified 25 rice stalks with at most two or three errors, whereas the men could hardly get two or more correct answers. These women not only described the rice, but also mentioned several other features, such as the ease with which husk can be removed, the length of time required to cook and suitability to different ecological conditions (Shiva *et al.*, 1990).

Since – as will be described in the next Chapter – hills/ mountains are the rich repositories of biodiversity, women's role in seed selection and vegetative propagation is essential both for agricultural practices as well as for biodiversity conservation.

In a small sample participatory study with women hill farmers in Dehradun, we were provided with not less than 145 species of forest plants that women have knowledge of and which they utilise. Hill and mountain economy is one in which forests, food crops, and livestock management are closely linked through biomass flows. Women's role in keeping these linkages and biomass or resource flows going is essential. It is through these linkages that ecological stability and sustainability is maintained.

The labour inputs, largely of women, make the resource flows possible, as she carries the fodder, fuel, and other minor forest products from the forest to the animals and the households; she brings the animals and the dung to the fields and feeds the animals with crop residues and the family with the field crops.

So the most critical input in the system, which provides the woman and her family with all the essentials (food, fodder, fuel, medicines, etc.) to fulfil their basic needs, is woman's work and knowledge. In fact, women in these hills are the wheels of the vehicle of economy, as Sajwan (1986) concludes.

When we try to analyse this situation according to the concept of autonomy, the following picture becomes clearer.

- ☆ In the traditional agro-ecosystem of the hill regions women's access to and control over means of production, such as forests, land and land use, crops, animals, other natural resources, and ecosystems is considerable;
- ☆ Her access to and control over the form of training and education needed in that particular environment is also present, as the knowledge system on the environment are transferred from mother to daughter, generation after generation;
- ☆ With regard to women's access to and control over her own labour, it is clear that she has control over her own labour (although it is merely a necessity to put in her time in the various activities) and that of her children; not over

that of her husband; to what extent patriarchal control over women's time plays a role here is not yet clear;

- And with regard to access to and control over income, it is obvious that in an intact hill agro-ecosystem monitory income is not a condition for survival at all; it is the productivity of the environment which supplies the women and their families with a form of natural income;
- About the forms of cooperation and organisation of the women which will enable and help them, to control their own affairs, it is well-known that women (and girls) often do much of their collection and agricultural work together; it is usually a small group of women or girls which goes out to the forests or the springs; a fact is that the women in the Himalayas, especially in the Garhwal region, have shown a very eminent picture of cooperation and organisation in their collective effort to prevent the felling of the forest: the Chipko Movement;
- About the (traditional) control by the women over her own sexuality and fertility in the hill regions, we do not have much information; but what is obvious is that the population in these areas was stable for a long time (which might indicate traditional birth control measures, e.g., through herbs); on the other hand pregnancy and birth are a part of daily life here, not extraordinary states, that will interfere with the daily activities; women in their third trimester of pregnancy will be found still working in the field; the general practice in the Tehri district in Garhwal is to provide a mother and child with a straw-bed in the cattle shed for as long as twenty-one days as the mother is considered 'achhut' (unclean); girls are married off at an age of 14 to 17 years; and
- Finally with regard to women's sense of dignity and self-respect and their right to self-determination – as related to gender conceptions – there might be an indication in their cultural expressions in their songs and dances and resistance to modernisation and commercialisation processes; they not only will then lose their control over resources, but also their sense of dignity (their knowledge becomes worthless) and self-respect. What is also striking

is the fact that women are usually very much aware of the conflict in their interests and the interests of their husbands (*e.g.*, resistance against fruit trees as compared to fuel-fodder trees). The heritage system in the Garhwal region is patrilinear.

Our conclusion is that in the traditional hill agro-ecosystem of the Garhwal Himalayas, the autonomy of the women is/was quite obvious.

The Impact of External Influences

If there must be a war, let the weapons be your healing hands, the hands of the world's women in defence of the environment. Let your call to battle be a song for the earth.

*–**Mostafa Tolba***

However, recent developments, mainly during the last five decades – but already starting with Western colonisation – have deteriorated the local systems and therewith the autonomy and position of women.

Already in several studies, such as in *Staying Alive* of Shiva (1989) and *Women and Environment in the Third World* of Dankelman and Davidson (1988), it has been indicated how modernisation processes in cropping, forestry, animal husbandry, and water management have had a destructive effect on both women and their environment. In the context of this study, we will restrict ourselves to the situation in the above mentioned area, the mountain community in Central Himalayas, Particularly in Garhwal.

Recent developments have had serious effects on the sustainability of the agro-ecosystems in the mountain areas. In fact, when the system gets disturbed, this also has consequences on local women. Also here, with the disappearance of the forests, women have to walk longer distances on rugged terrain. A case study from the Chamba Block Tehri district in Garhwal Himalayas shows that a woman, on an average, has to walk 5460 km per year now to gather firewood. The degradation of the natural vegetation has put a lot of drudgery on them (Singh, 1998).

Whereas, as has been mentioned before, local women in the Garhwal region could list 145 species of trees and their uses, the "new" forestry experts, in contrast, were able to name only 20 species of pine and five species of *sesham* (ICIMOD, 1989). This indicates the differences in knowledge on local genetic resources of local inhabitants and so-called (external) experts. Another impact of resource scarcity is that women are spending increasing amounts of time in the collection of fodder. Often eight hours daily are spent in this task. Groups of women are forced to go on overnight treks to fetch fodder from forest areas outside of their communities. This also leads to new insecurities, such as the fear for sexual threats (ICIMOD, 1989).

Women's crucial role in agriculture is gradually diminished by the introduction of new agro-technology and crop varieties, which are aimed at the males. Women's role becomes more and more that of a labourer, losing her control over production and access to resources. In the Henwal Valley of Garhwal, for example, where the women used to grow indigenous varieties of rice, the new high-yielding varieties are completely directed at males and commercial interests. Dwarf varieties, for example, which are promoted in the Green Revolution, reduce the straw biomass for fodder and fertilizer and hence reduce biomass flows that ensure the sustainability of women's livelihoods in agriculture. In the rainfed agriculture animal waste and crop residues are essential for the sustainability of the soils. A reduction of straw biomass leads to a reduction of organic manure, thus contributing to declining soil fertility and reduced food production.

A greater incidence of women's suicides corresponds with those areas where the hardship level has reached critical limits, especially there where new crops have replaced local varieties and women's access to resources has diminished (ICIMOD, 1989). The replacement of local varieties with so-called new varieties creates resource scarcity in the farming system. The shift, for example, from local pulses to introduced soybean implies a shift from domestic to industrial food processing, displacing women from their resources (Singh, 1998). Much of the present agricultural research is concentrated on increasing the yield of only certain parts of a crop (Jiggins, 1984). The potato and mustard crops provide fresh leaf vegetables in the mountain diets. The HYVs of these crops do not.

Weeding is predominantly women's work. The fertilizer use, needed for the HYVs, has stimulated weed growth dramatically, increasing women's work burden again. The off-season vegetable farming for export, presently one of the most popular development strategies for hill agriculture, has had similar effects of reducing biomass inputs for sustainable management and increasing women's work. For example, the replacement of coarse grains, such as millet, by vegetables for export reduces not just the local food availability, but also seriously impairs the availability of fodder. A rough estimate shows that the cultivation of such crops on one hectare of land adds three hours to the daily work burden of women (ICIMOD, 1989; Singh 1998).

The dairy development programmes which were aimed at the marketing of milk, led to a monopolisation of fodder resources of the village commons by rich landlords and denial of free access of fodder to the poorer women. As a Haryana woman puts it: "Now I have to steal the grass for my buffalo and when the landlord catches me he beats me" (Shiva, 1988). Further on, more and more evidence shows that the women are unable to manage the crossbred animals, as their feed and other requirements are quite different from that of their indigenous cattle (Singh, 1998, 2000). Concentrate feed changes the texture and composition of cow dung, making it unsuitable for women's use in managing soil structure.

A study in Garhwal by Manjri Mehta reveals that the shift from subsistence to commercial agriculture, through the introduction of cash crops and the market economy, has led to a reduction in women's sphere of influence and an increasing dependence on men for extension services, purchase of seeds, and handling of tools and money, indebtedness is increasing (ICIMOD, 1989). It is, for example, a serious trend that instead of bride prizes now dowries are being paid in the hill region. These developments threaten the position of women in particular. Further, their collective action to cope with resource shortages has been gradually eroded away.

Indicative for the present situation is further the fact that as their work burden has increased, in certain areas in Garhwal women are already brought out of the cattle shed just eleven days after delivery, so that they can continue their work (Shiva *et al.*, 2005).

Conclusions

Sensitivity on the part of policy-makers and natural resource managers is a sine qua non *for balancing biodiversity conservation with local people's concerns.*

– *Egbert Pelinck*

The above cases have shown that the introduction of new agro-technologies within the context of industrial-commercial market systems, are based on a sectoral, fragmented, and reductionist approach. The market-integration has resulted in resource fragmentation. The traditional biomass flows, maintained by women, get disturbed and linkages between the different sectors of the system – food crops, forests, and livestock and their linkages – break down. Inputs and outputs become completely dependent on external markets and flows, whereby the inputs are not rightly adapted to the agro-ecosystem. This process means that:

- Women's control over the natural resources (including land, water, and species) and the flows in the agro-ecosystem (nutrients, etc.) gets lost;
- The replacement of local varieties/landraces and biodiversity through a one-dimensional commercial option in agricultural sciences and policies causes loss of essential sources and elements to meet women's (and their families') needs (e.g., for fuel, fodder, and minor forest products);
- Food availability gets reduced, that hits women first;
- The vulnerability of the system increases and the ecological deterioration it causes makes women's position very uncertain;
- They loose their control over labour as a result of changing labour structures and the increases in their work burdens;
- Not only the replacement of diversity by uniformity, but also actual loss of (control over and access to) land and forests (including grazing lands) – through commercialisation and felling, or introduction of cash crops (HYVs) – makes women's work more difficult and increases her work burden;

- The de-skilling and de-intellectualising of women, through ignoring their contributions in management, knowledge, and experience in the system, results in a loss of women's knowledge with regard to forestry, cropping, plant genetic resources, and animal husbandry; it also deteriorates her cultural expressions; and
- Women loose their status and decision-making power in the social system; the technology is based on gender conceptions which break down women's sense of dignity and self-respect and her right of self-destination (e.g., by an increase of her dependence on external actors and inputs).

Concluding, one could say that the introduction of new agro-technologies not only has resulted in class differentiation, but also in environmental degradation and gender differentiation. The introduction of the new technologies has led to a reduction of women's autonomy.

Women are the real custodians and managers of biodiversity and mistresses of agro-technology. Rather than compartmentalization of agriculture into mere cropping, or animal husbandry, or horticulture etc., women look at and analyse agriculture as sum total of all land-related activities, all the processes through which soil fertility is maintained, conservation of biodiversity is ensured, balance amongst all components of farming (forests, pastures, cropland, livestock, etc.) is maintained, all organic linkages and processes within the farming system components are ensured, cyclic flows of nutrients are regulated, ecological integrity of farming systems is ascertained, diversity of foods is produced, small-scale rural industries based on biodiversity are run, and agro-ecophilosophy guiding the farming cultures to adopt all measures leading to transcendental changes and sustainability becomes the way of life. Women folk, through agriculture, maintain sanctity of nature and life.

References

Baer, Dagmar. 1988. *Women's Participation in off-farm Employment: Current Opportunities and Possible Options in Nepal*. Kathmandu: ICIMOD.

Dankelman, Irene and Davidson, Joan. 1988. *Women and Environment in the Third World: Alliance for the Future*. London: Earthscan.

Hoskins, Marilyn. 1979. *Women in Forestry for Local Community Development*. Washington DC: USAID.

ICIMOD. 1989. *Women in Mountain Development: Report of the International Workshop on Women, Development and Mountain Resources: Approaches to Internationalising Gender Perspectives*, Kathmandu, Nepal, 21-24 November 1988. Kathmandu: ICIMOD.

Sajwan, G.S. 1986. Socio-Economic Role of Women in Garhwal. Himalaya: *Man and Nature*, September 1986, 14-18.

Scrijvers, Joke. 1986. *Mothers for Life*. Eburon, Delft.

Shiva, Vandana. 1988. Women as Resource Integrators in Mountain Agriculture. Paper in ICIMOD Seminar (unpublished).

Shiva, Vandana. 1989. *Staying Alive: Women, Ecology and Survival in India*. New Delhi: Kali for Women.

Shiva, Vandana. 1990. Biodiversity Conservationn: Ecological vs Commercial Approaches. *The Ecologist*, Vol. 20, Nno. 2, 44-47.

Shiva, Vandana; Singh, V.; Dankelman, Irene, Negi, Biju and Singh, Shilpy. 2005 *Biodiversity, Gender and Technology in Mountain Agriculture: Glimpses of Central Himalayas, India.* New Delhi: Navdanya.

Singh, V. 1987. Hills of Hardship. *Hindustan Times*, New Delhi, January 18, 1987.

Singh, V. 1998. *Draught Animal Power in Sustainable Development of Mountain Agriculture: A Study of Perspectives and Issues in Central Himalayas, India*. Kathmandu: ICIMOD.

Chapter 10

Marine Species Under the Spell of Climate Change

☆ *Amrita Nag and Pooja P. Patel*

The world's seas are one of the biggest sinks or absorbers of carbon dioxide. As the amount of carbon dioxide in the atmosphere increases, the ocean over time begins to absorb more carbon dioxide as well. This has a direct effect on the ocean currents which flow around the world distributing salt, oxygen and heat. The currents have a large impact on the systems that live in the ocean, since many of the marine species rely on the currents for transportation, and the transfer of nutrients, temperature, and salinity. Rising temperatures have a profound impact on marine life – and not just for polar bears. Temperature rises are impacting on the entire marine food web. Phytoplankton, that underpin all life in the oceans, are likely to be destroyed by global warming. Since the breaking of the nourishment chain in nature once will lead to incredible results, the extinction of some species will directly affect the other species. For example, phytoplankton, which feeds small crustaceans including krill, grows under sea ice. A reduction in sea ice implies a reduction in krill; and krill feeds many whale species, including the great whales. Even Coral reefs have proved to be particularly sensitive to warming. In this paper the effect of global warming on the ocean currents is discussed and further an attempt has been made to explain the effect of global warming on marine species like Phytoplankton, Crustaceans, Cetaceans, some species of fishes, turtles and coral reefs.

Effect of Global Warming on Marine Species

Greenhouse Effect

The Earth is warmed thanks to sunlight that reaches earth's surface. As the earth's surface is much cooler than the sun, it radiates energy at much longer wavelength than the sun. Some of these longer wavelengths are absorbed by greenhouse gases in the atmosphere before they are lost to space. The absorption of this long wave radiant energy warms the atmosphere. The greenhouse gases also emit long wave radiation both upward to space and downward to the surface. The downward part of this long wave radiation emitted by the atmosphere is the greenhouse effect. Major Greenhouse gases are water vapor, carbon dioxide, methane, ozone, nitrous oxide, and CFCs. With the temperature control of the greenhouse effect, the Earth has an average surface temperature of 59°F (15°C). Without it, the average surface temperature would be 0°F (-18°C), a temperature so low that the Earth would be frozen and could not sustain life.

Global Warming

Greenhouse gases prevent infrared radiation from escaping into space, and this greenhouse effect maintains the earth's warm temperature. Scientists have concluded that human activities are contributing to global warming by adding large amounts of heat-trapping gases to the atmosphere. Our fossil fuel use is the main source of these gases. Every time we drive a car, use electricity from coal-fired power plants, or heat our homes with oil or natural gas, we release carbon dioxide and other heat-trapping gases into the air. The second most important source of greenhouse gases is deforestation, mainly in the tropics, and other land-use changes. Since pre-industrial times, the atmospheric concentration of carbon dioxide has increased by 31 percent. Over the same period, atmospheric methane has risen by 151 percent, mostly from agricultural activities like growing rice and raising cattle. As the concentration of these gases grows, more heat is trapped by the atmosphere and less escapes back into space. An increase in global temperature can in turn cause other changes, including a rising sea level and changes in the amount of precipitation. These changes may increase the frequency and intensity of extreme weather events, such as floods, droughts, heat waves, hurricanes and tornados. Higher or lower agricultural yields, glacier retreat, reduced summer stream flows, species extinction and increase in the ranges of disease vectors are the other consequences.

United Nations Environment Programme (UNEP) and the World Meteorological Organization (WMO) in 1988 set up the **Intergovernmental Panel on Climate Change (IPCC)** to examine the most current scientific information on global warming and climate change. The IPCC concluded in its Third Assessment Report, "An increasing body of observations gives a collective picture of a warming world and other changes in the climate system." The kinds of changes already observed that create this consistent picture include the following:

Examples of observed climatic changes, as elaborated in reports of IPCC, are:

- ✰ Increase in global average surface temperature of about 1°F in the 20th century;
- ✰ Decrease of snow cover and sea ice extent and the retreat of mountain glaciers in the later half of the 20th century;
- ✰ Rise in global average sea level and the increase in ocean water temperatures;
- ✰ Likely increase in average precipitation over the middle and high latitudes of the Northern Hemisphere, and over tropical land areas;
- ✰ Increase in the frequency of extreme precipitation events in some regions of the world.

Examples of observed physical and ecological changes are:

- ✰ Thawing of permafrost;
- ✰ Lengthening of the growing season in middle and high latitudes;

- ☆ Poleward and upward shift of plant and animal ranges;
- ☆ Decline of some plant and animal species;
- ☆ Earlier flowering of trees;
- ☆ Earlier emergence of insects;
- ☆ Earlier egg-laying in birds.

Global Warming and Ocean Currents

The ocean currents include the mass movements and circulations of water caused by various factors. They flow throughout the world connecting the oceans water and resources. The patterns of these movements are determined by wind, salinity, heat distribution, major topographic landmarks, and the rotation of the earth. Ocean currents have two different methods of moving water: (i) surface currents move as a result of blowing winds and are influenced by atmospheric factors driven by forces such as tides, wind and gravity. Currents keep oceans in constant motion; and (ii) the deep ocean currents commonly referred to as the oceans conveyor belt move as a result of the density differences that are present in the ocean.

Density is determined by both salinity and temperature, so in areas such as the Poles the water is very cold and since glaciers are forming freshwater is taken out of the water and the remaining water becomes more saline resulting in very dense water. Hence water dense enough to sink to the ocean floor is formed in Polar Regions by surface water freezing, which concentrates salt in very cold water beneath the ice. The dense water then sinks. This water that is formed in the poles then moves throughout the world mixing the oceans. Conversely, warmer and less dense water moves towards the poles from the equator, where it is converted into water to be used in thermohaline circulation. The ocean flows around the world distributing salt, oxygen and heat. This is called "thermohaline.

The two largest currents on earth are the Kuroshio Current which circulates near Japan, and the Antarctic Circumpolar Current, which circulates around Antarctica. The impact of global warming on the vast Southern Ocean around Antarctica is starting to pose a threat to ocean currents that distribute heat around the world. Melting ice-sheets and glaciers in Antarctica are releasing fresh water, interfering with the formation of dense "bottom water," which sinks 4-5 kilometers to the ocean floor and helps drive the world's ocean circulation system. A slowdown in the system known as "overturning circulation" would affect the way the ocean, which absorbs 85 percent of atmospheric heat, carries heat around the globe.

Global Warming and Marine Species

"There has long been a belief that the sea, at least, was inviolate, beyond man's ability to change and to despoil. But this belief, unfortunately, has proved to be naive."

–*Rachel Carson*

The world's seas are one of the biggest sinks or absorbers of carbon dioxide. As the amount of carbon dioxide in the atmosphere increases, then over time the ocean begins to absorb more carbon dioxide as well. The problem is that it makes the water more acidic which is not good for anything made from calcium carbonate which includes corals and plankton, as they are easily dissolved by acid. Dissolved carbon dioxide could increase the acidity of sea water by as much as 0.5pH units. Such a change would disturb the oceans' chemical balance and kill off marine life. Since the beginning of the industrial revolution human have pumped an estimated total of 450 billion tones of carbon dioxide into the atmosphere, around half of which has ended up in the ocean.

Dr. Stefanie Kaiser is a German scientist who specializes in studying small seabed animals. She said, "Although many of these animals are tiny, their behaviour helps us paint a much bigger picture in terms of how marine life may react to changes to the environment." Rising temperatures have a profound impact on marine life – and not just on polar bears. The NOAA "Climate Change" article explains how marine creatures are adapted to certain temperatures and as temperatures change species will migrate to the temperatures they are suited for. Those that can't migrate fast enough, for example clams, could die out. The September 2, 2009 Associated Press article "Change is seen in Atlantic from climate, fishing" says that new population distributions have already been noticed off the United State's East Coast due to climate change and human fishing practices.

Recently, a sign of climate change was seen in Canada's Gulf of St. Lawrence. There was not enough ice to provide necessary habitat for thousands of baby harp seals, and nearly 100 per cent of all pups born this year drowned.

Effect of Global Warming on Coral Reefs

Shallow ocean habitats that are rich in life and these are inhabited by coral reefs. The reef's massive structure is formed from coral polyps, tiny animals that live in colonies. When coral polyps die, they leave behind a hard, stony, branching structure made of limestone. The coral provides shelter for many animals in this complex habitat, including sponges, nudibranchs, fish (like Blacktip Reef Sharks, groupers, clown fish, eels, parrotfish, snapper, and scorpion fish), jellyfish, anemones, sea stars (including the destructive Crown of Thorns), crustaceans (like crabs, shrimp, and lobsters), turtles, sea snakes, snails, and mollusks (like octopuses, nautilus, and clams). Birds also feast on coral reef animals. The Great Barrier Reef (off the coast of NE Australia) is the largest coral reef in the world. It is over 1,257 miles (2000 km) long.

Threat of existence is looming large on the coral reefs. Global warming is one of the causes. They can only live in waters between 18°C and 30°C, an unprecedented increase in the number of coral bleaching events during the past 2 decades (which have had some of the warmest years in history). When ocean temperatures get too high, coral polyps lose the symbiotic algae inside them, causing them to turn white, or "bleach," and eventually die. Coral bleaching is the whitening of coral colonies due to the loss of symbiotic zooxanthellae from the tissues of polyps. This loss exposes the white calcium carbonate skeletons of the coral colony. According to Dr. Ove-Hoegh Guldberg, coral bleaching occurs when algae in a symbiotic relationship with the coral become dysfunctional and expelled. The algae (dinoflagellates) normally provide foods for the coral through photosynthesis. However, when the temperature increases, the ability to perform photosynthesis is completely shut down.

Warmer Seas – Threat to Phytoplankton

The plants are known as phytoplankton. They lie at the base of the marine food chain because they convert sunlight and carbon dioxide into organic carbon–food for everything else. The nutrients, containing nitrogen, phosphorus and iron, are vital food for phytoplankton. If the supply is interrupted the plants die off, which prevents them from absorbing carbon dioxide from the atmosphere.

The vast quantities of phytoplankton in the oceans absorb the greenhouse gas carbon dioxide during photosynthesis and when the organisms die they fall to the seabed, carrying their store of carbon with them, where it stays for many years thereby helping to counter global warming. Phytoplankton therefore acts as a carbon "sink" which takes carbon dioxide from the atmosphere and deposits the carbon in long-term stores that can remain undisturbed for thousands of years. Eventually this results in reducing the carbon dioxide level in the atmosphere. If the growth of phytoplankton is negatively affected by global warming, this ability to act as a shield against global warming is also affected leading to a much-feared positive feedback.

In the 2007 report of Intergovernmental Panel of Climate Change, it is stated that if the average temperature of the earth increases by 1-2°C, one third of the species on the earth will change their distribution areas or become extinct. Anchovy (*Engraulis encrasicholus*) and sprat (*Sprattus sprattus*) are fishes which feed on plankton and they make the organic loads absorption possible and if this doesn't happen and the plankton drifts in the bottom. H_2S layer will rise even more.

Crustaceans

An increase in the amount of CO_2 leads to a decline in pH due to which the sea water becomes more acidic. The external skeleton of crustaceans is made of a common form of calcium carbonate and this can liquefy in the acidic water. The acidic water unconstructively affects the crustaceans. If these crustaceans which are at the bottom of the nutrition chain extinct then the entire ecosystem will change. Winter ice is of great importance for algae and small crustaceans that are called krill which feed on algae. In the last 30years there is a decline in the population of Krill by 80 per cent on average and the reason behind this is the decreasing of the ice cover carried by South Ocean in winter.

Whales, Dolphins and Porpoises (Cetaceans)

Because the impact of global warming is currently greatest at the Earth's poles, cetaceans that rely on polar, icy waters for their habitat and food, such as belugas, narwhals (the creatures of unicorn legends) and bowhead whales, are likely to suffer the most. Here's why: Krill, tiny shrimp-like animals that are a major food source of whales, live under Antarctic sea ice during the winter. As the sea ice cover has diminished, scientists have observed massive decline in krill populations, threatening the whales' food supply. Global warming could also threaten cetaceans through warmer ocean temperatures, the acidification of the ocean as it absorbs the increasing amounts of carbon dioxide in the atmosphere, increased susceptibility to disease and reduced reproductive success. The great whales also risk losing their feeding grounds, in the Southern Ocean around Antarctica, because of the melting and collapse of ice shelves. Whole species of marine animals and fish are directly at risk due to the temperature rise; they simply cannot survive in warmer waters. Some penguin populations, for example, have decreased by 33 percent in parts of Antarctica, because of habitat decline. An increasing occurrence of disease in marine animals is also linked to rising ocean temperatures.

Turtles

In most living beings the chromosomes decide the sex of juveniles but in some living beings such as tortoises it is the temperature of the habitat that determines whether the juvenile will be male or female. Male juveniles are formed in lower temperatures (20-27°C), and female juveniles are formed in temperatures higher than 30°C. It can be a male or female if the temperature is between 28–31°C. Even in alligators the temperature determines the sex of the juvenile. A study was carried out in USA and it was found that because of an increase of 1°C in the habitat temperature nearly all the juvenile turtles were females. Hence the increase in temperature is detrimental for the population of turtles as the new juveniles will only be females and hence the sex ratio will be imbalanced in future (Anonim, 2008d).

New studies have found that Pacific salmon, especially sockeye, are particularly vulnerable to temperature changes which affect their metabolism. The warmer the water, the more food they need to stay alive. In 1997-1998, higher sea temperatures during the winter may have led to the collapse of western Alaskan salmon populations, according to the report which concluded that the Pacific sockeye and other salmon species may soon be at risk of extinction.

Conclusion

1. Global warming looming large on the vast Southern Ocean around Antarctica is starting to pose a threat to ocean currents that distribute heat around the world. Melting

ice-sheets and glaciers in Antarctica are releasing fresh water, interfering with the formation of dense "bottom water," which sinks 4-5 kilometers to the ocean floor and helps drive the world's ocean circulation system. A slowdown in the system known as "overturning circulation" would affect the way the ocean, which absorbs 85 percent of atmospheric heat, carries heat around the globe.

2. The whole marine food web is bound to be severely affected by global warming. For example, phytoplankton, which feeds small crustaceans including krill, grow under sea ice. A reduction in sea ice implies a reduction in krill; and krill feeds many whale species, including the great whales.
3. Global warming is particularly posing threat to some sensitive life organizations, *e.g.*, the coral reef. As surface temperatures have risen in recent years, many reefs have bleached – meaning they expel the colourful algae that produce the foods on which they rely – and, if temperatures remain too high for too long and bleaching persists, the corals die.
4. If the temperature rising trend goes on unabatedly, it might knock down numerous marine species.

References

commondreams.org

http://dergiler.ankara.edu.tr/dergiler/47/1037/12515.pdf

http://marinebiologyoceanography.suite101.com/article.cfm/how_global_warming_affects_the_ocean

http://news.mongabay.com/2008/0217-acidification.html

http://sitemaker.umich.edu/section2group2/background_information_on_currents

www.aims.gov.au/pages/research

www.defenders.org

www.livescience.com/environmentoceanworld.tamu.edu

www.oceans.greenpeace.org

www.sciencedaily.com

www.stopwhaling.org

www.ucsusa.org/global_warming/ess.geology.ufl.edunews.nationalgeographic.com

Part III

Biodiversity Restoration: Innovative Management Approaches

their genetic plasticity and degree of specialization. Specialists adapted to local conditions will be more likely to go extinct than generalists (Allan *et al.*, 2005).

Because global climate change will shift the ranges of temperate fishes towards higher latitudes, it could have potentially serious impacts on some groups of fishes: fishes in geographically isolated environments and fish in lentic systems. Migration towards the poles may be an option for some species, but fishes in east–west oriented stream systems, e.g., the southwestern United States, eastern Africa (Allan *et al.*, 2005), and New Zealand would be prevented from moving to higher latitudes by excessive distance and physiological barriers like the ocean. Though some fishes in streams with an east-west orientation could migrate toward high-elevation headwaters in search of cooler temperatures, they would likely be faced with an entirely new set of environmental challenges such as smaller stream size and unfamiliar habitat (e.g., higher gradient, larger substrate, changes in turbidity). Similarly, fishes forced to seek optimal temperatures in north–south oriented systems would probably also have to cope with a new physical environment.

Fishes in relatively closed systems such as lakes will also be affected with potentially detrimental results. Fisheries managers could alleviate this phenomenon by translocating "trapped" fishes into more suitable watersheds (Shuter *et al.*, 2002). The introduction of fish to new environments is not always successful and raises environmental, political, and ethical questions (Keeton, 2002). When various Colorado subspecies of cutthroat trout (*Oncorhynchus clarki*) were introduced into fishless or reclaimed waters, they experienced a success rate of less than 50 per cent. The introduction of fish into naive waters is not a decision to be made lightly because of the possible negative consequences for organisms already in that environment. History has taught us that introduction of non-indigenous species into natural ecosystems can have disastrous results (Courtenay, 1997). For example, the Nile perch (*Lates niloticus*) was introduced into Lake Victoria in a well-meaning effort to improve the fishery. The endemic cichlids had not evolved with a large, cursorial predator and were subsequently decimated.

Possible Solutions

The future impacts of climate change on fisheries and aquaculture are still poorly understood. The key to minimizing

Chapter 11

Agroforestry Systems: An Alternative Avenue for Livelihood Security and Mitigating Climate Changes in the Barind Tract of Bangladesh

☆ *M.S. Bari and M.A. Rahim*

This chapter summarizes the current knowledge on Agroforestry practices in the Barind tract of Bangladesh. The part of greater Rajshahi, Dinajpur, Rangpur and Bogra district of Bangladesh is geographically identified as the Barind tract. The hard red soil of these areas is very significant in comparison to that of the other parts of the country. A typical dry climate with comparatively high temperature prevails in the Barind area. Scanty rainfall and excessive extraction of groundwater has created an alarming situation among the Barind people and others about irrigation to the rice fields. In addition, the situation may cause an adverse impact on the people whose livelihood largely depends on agriculture. The groundwater table has been declined by at least 10 to 15 meters in some areas of the Barind tract during the last 20 years. Due to absence of forestland in the Barind tract, the soil organic carbon content can decrease more rapidly because of the higher rate of decomposition. The Barind soils are largely carbon depleted but can be brought back to carbon-carrying capacity by bringing under suitable Agroforestry systems. Agroforestry has tremendous potential for carbon sequestration not only in above C biomass but also root C biomass in deeper soil depths. Agroforestry systems habitually conserve soil organic carbon and maintain a great natural nutrient pool. Cultivation of trees with arable crops under Agroforestry systems help in improving soil organic carbon content and sequestration carbon particularly in the highly degraded Barind lands. Agroforestry systems are a way of securing the growth of cash crops that may constitute an alternative source of income in moments of crisis. Besides being a source of fuel wood, a greater presence of trees in cropping system contributes to decreasing temperatures and to increasing rainfall, thus contrasting the negative environmental impact of climate changes. Agroforestry systems have generated a lot of employment opportunities in the Barind area. More crops mean involvement of more people in various activities like involment in dairying, sericulture, apiculture and additional associated agro-based interventions. Successful adoption of Agroforestry practices in the Barind area have shown that the Agroforestry practitioners of this area were very sound positioned economically, and had added social status too.

Introduction

The impact of climate variability and change on farming production is a worldwide distress. However, the impact is on the whole significant in Bangladesh where agriculture is the leading sector of the Bangladesh economy, accounting for some 35 percent of the GDP and 63 percent of the labor force (BBS, 2008). Agricultural production is already under pressure from increasing demands for food and the corresponding problem of running down land and water resources caused by overuse and pollution. Impacts of climate variability and change cause a supplementary risk for agriculture. Global circulation model (GCM) results predict an average temperature increase in Bangladesh due to climate change of 1.0°C by 2030 and 1.4°C by 2050. Though monsoon precipitation is likely to increase by 6.8 percent by 2050, the distribution patterns of precipitation during the growing season, high temperature and higher rates of evapotranspiration will create further water stress conditions and a decline in agricultural production in the drought-prone areas like the Barind tract of Bangladesh. The part of greater Rajshahi, Dinajpur, Rangpur and Bogra district of Bangladesh is geographically identified as the Barind tract. A continued trend of more frequent and intense droughts, as a result of further climate variability and climate change, is anticipated to have considerable impacts on the agricultural sector of the Barind region. The hard red soil of these areas is very significant in comparison to that of the other parts of the country. A typical dry climate with comparatively high temperature prevails in the Barind area. Scanty rainfall and excessive extraction of groundwater has created an alarming situation among the Barind people and others about irrigation to the rice field. In addition, the situation may cause an adverse impact on the people whose livelihood largely depends on agriculture.

The groundwater table has been declined by at least 10 to 15 meters in some areas of the Barind tract during the last 20 years. Due to absence of forestland in the Barind tract, the soil organic carbon content can decrease more rapidly because of the higher rate of decomposition. The existing land use systems in this area will become more vulnerable owing to increase in the atmospheric temperature, levels of CO_2 and other greenhouse gases. The result would be drastic reduction in productivity potential of the system. Nevertheless, the limits of agricultural production even using the most intensive high input agriculture have already been reached. The scenario on fuel, fodder and timber production front is also not reasonable. So, the situation is very alarming and under these upsetting situations, adaptation to climate change is one of the approaches considered likely to reduce the impacts of long-term changes in climate variables. Adaptation is a process by which strategies to moderate and cope with the consequences of climate change, including climate variability, can be enhanced, developed and implemented (UNDP, 2004). So, the efficient management of natural resources is the call for of the hour. The existing land use systems with separate allocation to agriculture and forest are insufficient to meet the demands for food, fuel, fodder, timber and other minor products in the 21st century. One should follow effective and compatible cultivation approaches where fruits, vegetables, spices, medicinal plants and timber can be grown combined in the limited land. In this link, the agroforestry system may be the best alternate cultivation approach. By practicing this cultivation arrangement, one can efficiently intensify the production of fruits, vegetables, spices, medicinal plants and timber vertically. Consequently, agroforestry is considered a universal remedy for overcoming most of the problems related to the alleviation of poverty, socio-economic insecurity and shrinking ill effects of the climate change. Therefore, we should follow a line of investigation to formulate viable, environment friendly and widely accepted agroforestry systems. On the other hand, there is an urgent need to investigate the potentials of growing forest and fruit tree species with high valued cash crops like vegetables, spices and medicinal plants on the woodlots/orchards to attain predictable understanding that will have wide applicability, economic feasibility and suitability for the conditions of Bangladesh especially for the Barind tract. In fact, there is practically no information available on the potentiality of growing vegetables, spices and medicinal plants with multipurpose tree species except for the exploratory studies made on inclusion of some research in India, which showed improved productivity (Pareek and Gupta, 1984; Maheshwari *et al.*, 1985). Considering all these facts, the proposed study was undertaken to evaluate the role of agroforestry systems on the improvement of the soil properties and the livelihood security systems under the area of the Barind Tract of Bangladesh conditions.

Materials and Methods

In order to fulfill the objectives of this study, two experiments were conducted. The first experiment was survey on the impact of existing agroforestry system on the livelihood security in the Barind tract of Bangladesh and the second one was the role of agroforestry system on the improvement of soil properties in a multilayered coconut orchard.

For the first experiment, survey was conducted in the year 2009 on 90 farms in 9 unions of three upazila namely, Atrai, Boraigram and Puthia were selected from the district of Naugaon, Natofe and Rajshahi, respectively. The villages and upazilas were selected considering the representativeness of the Barind tract of Bangladesh. At first the farmers who have a minimum of two horizontal strata that provided part of the daily nutritional needs of the families were listed. Out of 90 respondent farmers, 30 were selected from each upazila. From each upazila three unions were randomly selected and from each union 10 respondent farmers for the survey were selected based on four different farm categories. Information was collected through a structured pre-tested questionnaire by interviewing the head of the farm and by direct observation. Simple and direct questions and some scales were included in the schedule to collect data for the study. Data collected from the respondents (samples) were verified, compiled, tabulated and analyzed statistically according to the objectives of the study. In some cases, qualitative data were converted into quantitative data by means of suitable scoring to facilitate interpretation. Considering the farmer's behaviours (observed) conditions and limitations in respect of time, money and other resources available to the researcher, ten characteristics of the farmers selected as the independent variables for this study. The selected variables included (i) age, (ii) education, (iii) family size, (iv) farm size, (v) annual income, (vi) knowledge of agroforestry production system, (vii) communication exposure, and (viii) problem confrontation. However, knowledge of agroforestry farming refers to the ability of respondent to recall or recognize items of information related to agroforestry production system. It was measured by computing an agroforestry knowledge score. Each respondent was asked to answer all the 11 questions. A respondent could get a score 2 against each item for correct response, l for partial response, and 0 for wrong response. Thus agroforestry knowledge of respondent could range from 0 to 50, zero (0) indicating no agroforestry knowledge and 50 indicating very high agroforestry knowledge. In case of the measurement of problem confrontation, it was measured in one way such as, using of closed form of questions. The farmers were asked to give their opinion on 15 selected problems, which were identified during pre-testing of the questionnaire along with their extent of confrontation in use of agroforestry production system. A four point Likert scale was used for computing the problem confrontation score. The weights assigned were 3 for "high", 2 for "medium", 1 for "low" and 0 for "not at all". The problem confrontation score of the farmers could range from 0 to 48; zero indicating no problem and 48 indicating high problem confrontation. However, farmers' opinion on the improvement of socio-economic condition as well as local environment through agroforestry production system was the dependent variable in this study. A three point Likert scale was constructed for measuring farmers' opinion on agroforestry production system. A farmer was asked to express his extent of agreement or disagreement in respect of each of the 10 items. Each respondent expressed his views, which was placed against any of the three responses. Scoring was made in the following manner for positive or negative statements.

Table 11.1: Salient Features of the Selected Characteristics of the Respondents (N = 90)

Responses	*Weights for Positive Statement*
Agree	3
No opinion	2
Disagree	1

The second experiment was conducted in the existing multistoried coconut garden of the Department of Horticulture, Bangladesh Agricultural University, Mymensingh throughout the growing seasons of 2005-2008. The experimental area was under sub-tropical climate characterized by heavy rainfall during the months from April to September and scanty rainfall during the rest period of the year. The soil of the experimental area was silty loam in texture. It was a medium high land and fertile well drained. The experiment was laid out in randomized complete block design with five replications. Treatments were: T_1- comprises the combination of coconut + guava + turmeric based agroforestry system, T_2- comprises the combination of coconut + lemon + turmeric based agroforestry system, T_3- individual turmeric plant in open condition. In the open condition T_3 received 100 per cent sunlight; while coconut + guava based multistoried agroforestry system (T_1) and coconut + lemon based multistoried agroforestry system (T_2) allowed, respectively 52 to 56 per cent and 63 to 68 per cent sunlight for the growth of the turmeric plants. The coconut trees were planted thirty years before and the spacing of plantation was 8m x 8m. As the middle or second layer plant, guava and lemon both were five years old and were in full bearing condition. Spacing between guava and/or lemon and between rows was 3m × 3m. Coconut, guava and lemon trees were not dressed with fertilizer. But, the ground layer crop, *i.e.* turmeric, was fertilized with the recommended doses of fertilizers for the each growing season. The experimental plots were kept weeds free by weeding frequently. The plots were irrigated whenever needed by using hose pipe and watering cane. Soil samples were collected at 15cm depth of 5 replications. Soil samples were sieved for laboratory analysis for soil pH, organic carbon, total nitrogen, available phosphorus and exchangeable potassium (Gillman and Sumpter, 1986; Menzies and Gillman, 1997). Soil data is used to compare between agroforestry practices and sole cropping of coconut plantation.

Results and Discussion

Experiment No.1: Survey on the Impact of Existing Agroforestry System on the Livelihood Security in the Barind Tract of Bangladesh

Selected Characteristics of the Respondents

According to the objectives of the study, data were collected from a sample of 90 respondents. The salient findings of seven selected socioeconomic characteristics of the respondents have been presented in Table 11.2.

Age

Age of the respondents ranged from 25 to 62 years. On the basis of age, the farmers were classified into three categories, as young, middle aged and old. Data presented in Table 11.2 indicated that the highest proportion (40.0 percent) of the farmers was in the middle aged category compared to 31.11 percent young aged and 28.89 percent old category. It may also be revealed that middle and young aged (70 percent of the respondent) who were found to be more interested, energetic and enthusiastic for establishing agroforestry production system. Particularly the middle aged people are well experienced. Thus decision-making regarding the agroforestry practices in the study area might be influenced by age category of the rural farmers.

Education

The level of education score of the respondents ranged from 0 to 12; the average being 2.40 and the standard deviation was 3.26. Based on their level of education, the farmers were grouped into five categories and the large proportion (42.22 percent) of the farmer fell under the category "can sign only" compared to 26.67 percent with primary education, 15.56 percent having no education, 11.11 percent having secondary education and to 4.4 percent having above

Table 11.2: Salient Features of the Selected Characteristics of the Respondents (N = 90)

Sl.No.	Characteristics Measuring Unit	Categories	Respondent (No.)	Per cent	Average	Standard Deviation
1.	Age	Young age (20 to 35)	28	31.11	42.36	10.86
	(years)	Middle age (35 to 50)	36	40.00		
		Old age (above 50)	26	28.89		
2.	Education	No education level (0)	14	15.56	2.40	3.26
	(years of schooling)	Can sign only (0.5)	38	42.22		
		Primary level (1-5)	24	26.67		
		Secondary level (6-10)	10	11.11		
		Upper secondary level(11-12)	4	4.44		
3.	Family size	Small (up to 4)	16	17.78	6.44	2.65
	(number)	Medium (5-8)	54	60.00		
		Large (above 8)	20	22.22		
4.	Farm size	Marginal (up to 0.40 ha)	16	17.78	0.63	0.28
	(hectare)	Small (0.41 to 1.0 ha)	68	75.55		
		Medium (1.0 to 2.99 ha)	6	6.67		
		Large (above 3.0 ha)	0	0		

Contd...

Table 11.2–Contd...

Sl.No.	*Characteristics Measuring Unit*	*Categories*	*Respondent (No.)*	*Per cent*	*Average*	*Standard Deviation*
5.	Annual income	Marginal (up to 25)	14	15.55	54.93	20.79
	(taka)	Small (26 - 45)	18	20.00		
		Medium (45 - 75)	51	56.67		
		Large (76 and above)	7	7.78		
6.	Communication exposure	Low contact (up to 12)	16	17.78	13.78	2.58
	(days)	Medium contact (13 to 18)	69	76.67		
		High contact (above 18)	5	5.55		
7.	Knowledge on agroforestry	Low knowledge (0 to 15)	14	15.56	20.48	4.12
		Medium knowledge (16 to 25)	68	75.56		
		High knowledge (above 25)	8	8.88		

secondary education. The literacy rate (+15 years) of the country is 55 percent (BBS, 2008). Thus the findings indicate that in the study area, the literacy rate seems to be higher than the national level average.

Family Size

The family members of the farmers ranged from 2 to 14 with an average of 6.44 and standard deviation was 2.65. Based on the family size, the farmers were divided into three categories. The highest proportion (60.0 percent) of the respondents fell under the medium family size category compared to 17.78 percent small and to 22.22 percent large family size categories. These findings also indicated that 82 percent of the respondents had either medium or large family size. The average family size (6.44) of the respondents of the study area was too much higher than the national average of 5.40 (BBS 2008). This may be due to prevalence of joint family or lack of proper adoption of family planning measures among the respondents in the study 'area. Another reason was that the national average of 5.40 persons per family was concerned with the rural and urban families, but the present study is concerned with the rural families only.

Farm Size

Farm size of the respondents was found to vary from 0.12 to 1.37 hectare. The average farm size was 0.63 hectare with a standard deviation of 0.28. The respondents were classified into four categories. It was observed in Table 11.2 that the highest proportion (75.55 percent) of the farmers had small farm compared to 17.78 percent marginal, 6.67 percent medium and none were found as large farm. Similar findings also reported by Sarkar (1997).

Annual Income

Annual income was measured in thousand taka per year and scored 1 for each thousand taka. The range of annual income score was 22 to 120 with an average of 54.93 and standard deviation 20.79. On the basis of annual income, the respondents were divided into four categories and the.data revealed that the highest proportion (56.67 percent) of the respondent had high annual income compared to 20.00 percent had medium annual income, 15.55 percent had low annual income and 7.78 percent had large income. Thus the majority farmers had high annual income indicating that the farmers having comparatively higher economic standings usually practice agroforestry practices.

Communication Exposure

The computed extension media contact scores of the respondents ranged from 8 to 20 with an average 13.78 and standard deviation of 2.58 against the possible range of' 0 to 30. On the basis of' extension media contact scores, the respondent were classified into three categories like low contact, medium contact and high contact. It was observed in Table 11.2 that the highest portion (76.67 percent) of the rural farmer had medium extension media contact; while 17.78 percent had low extension media contact and 5.55 percent of the respondent had the high extension media contact.

Knowledge on Agroforestry Production System

The respondent's Knowledge on agroforestry farming raged from 12 to 26 with an average of 20.48 and standard deviation of 4.12. It was observed that the majority (75.56 percent) of the farmers had medium knowledge compared to 15.56 percent low knowledge and 8.88 percent having high knowledge. It reveals that the majority of the farmers in the study area were under medium knowledge to low knowledge categories.

Problems Confrontation of the Farmers in Establishment of Agroforestry Production System

Problem in each item has been presented with frequency distribution of the farmers in per cent. For clarity of understanding of problems of the farmers, an index for each item along with rank order was computed by using the following formula:

$$\text{Problem facing index (PFI)} = Ph\times3 + Pm\times2 + Pl \times 1 + Pn\times0$$

where,

Ph: No. of respondents with "high problem"

Pm: No. of respondents with "medium problem"

Pl: No. of respondents with "low problem"

Pn: No. of respondents with "not at all problem"

Per cent of distribution of the farmers according to their problem in each of the 15 items has been shown in Table 11.3 along with problem facing index (PFI) and rank order of each problem. PFI of the farmers of the 15 items in establishing agroforestry practices

Table 11.3: Problem Faced by the Respondents Related to the Agroforestry Production System

Problem	*Number of Citation*				*PFI*	*Rank Order*
	High	*Medium*	*Low*	*Not at All*		
Lack of capital	58x3	18x2	12x1	2x0	210	1
Lack of credit	50x3	20x2	18x1	2x0	208	2
Lack of technical knowledge on multistoried production system	47x3	21x2	20x1	2x0	203	3
Scarcity of space	43x3	26x2	21x1	0x0	202	4
Absence of verified particular multistoried model	41x3	29x2	16x1	4x0	197	5
Lack of knowledge about species compatibility	40x3	31x2	14x1	5x0	196	6
Above ground negative effect on ground layer crops	36x3	29x2	18x1	7x0	184	7
Difficulties in post-harvest operations	34x3	29x2	22x1	5x0	182	8
Risk of pest infestation	30x3	21x2	28x1	11x0	160	9
High price of quality propagates	27x3	23x2	31x1	9x0	158	10
Low market price	24x3	22x2	28x1	16x0	144	11
Lack of training on multistoried cropping	22x3	18x2	36x1	14x0	138	12
Unavailability of farming inputs in time	20x3	16x2	36x1	18x0	128	13
Damage crops by animals	15x3	17x2	36x1	22x0	115	14
Conflict with neighbors	9x3	12x2	44x1	25x0	95	15

ranged from 95 to 210 against a possible range of 0 to 270 according to the rank order.

Table 11.3 shows that lack of capital was the most severe problem faced by the rural farmers, the next four problems based on their descending order of severity were "Lack of credit", "Lack of technical knowledge on agroforestry production system", "Scarcity of space" and "Absence of verified particular agroforestry production model". The Table 11.3 also shows that the least severe problems were "Damage crops by animals" and "Conflict with neighbors".

Farmers' Opinion Regarding the Improvement of Livelihood Security and Local Environment through Agroforestry System

Farmers' opinion regarding the improvement of livelihood security and the local environment aspects due to practicing agroforestry production system have been presented with frequency distribution of the farmers in Table 11.4. For clarity of understanding of opinions of the farmers, an index for each item along with rank order was computed by using the following formula:

$$\text{Farmers Opinion Index (FOI)} = Pa \times 3 + Pn \times 2 + Pd \times 1$$

where,

Pa: No. of respondents with "Agree"

Pn: No. of respondents with "No opinion"

Pd: No. of- respondents with "Disagree"

Farmers Opinion Index (FOI) of the 10 items in the improvement of socio-economic as well as local environmental aspects due to practice agroforestry production system ranged from 214 to 262 against a possible range of 0 to 270. The 10 opinions are listed in Table 11.4.

Table 11.4 shows that Increase income was the most acceptable opinion by the rural farmers. The next four opinions based on their descending order of agreement were "Increase economic security", "improve social status", "Improve local environment" and "Increase the supply of timber and fodder". The Table 11.4 also shows that the two least acceptable opinions were "Increase employment opportunity" and "Decrease family malnutrition", respectively.

Table 11.4: Farmers' Opinion Index (FOI) for the improvement of Socio-Economic Condition Through Agroforestry Production System

Item-Wise Statement	*Agree*	*No opinion*	*Disagree*	*FOI*	*Rank order*
Increase income	82x3	8x2	0x1	262	1
Increase economic security	79x3	11x2	0x1	259	2
Improve social status	76x3	14x2	0x1	256	3
Improve local environment	75x3	15x2	0x1	255	4
Increase the supply of timber and fodder	74x3	15x2	1x1	252	5
Solves fuel problem	72x3	16x2	2x1	250	6
Increase the availability of fruits	68x3	18x2	4x1	244	7
Increase vegetable and spices production	65x3	14x2	11x1	234	8
Increase employment opportunity	58x3	15x2	17x1	221	9
Decrease family malnutrition	54x3	16x2	20x1	214	10

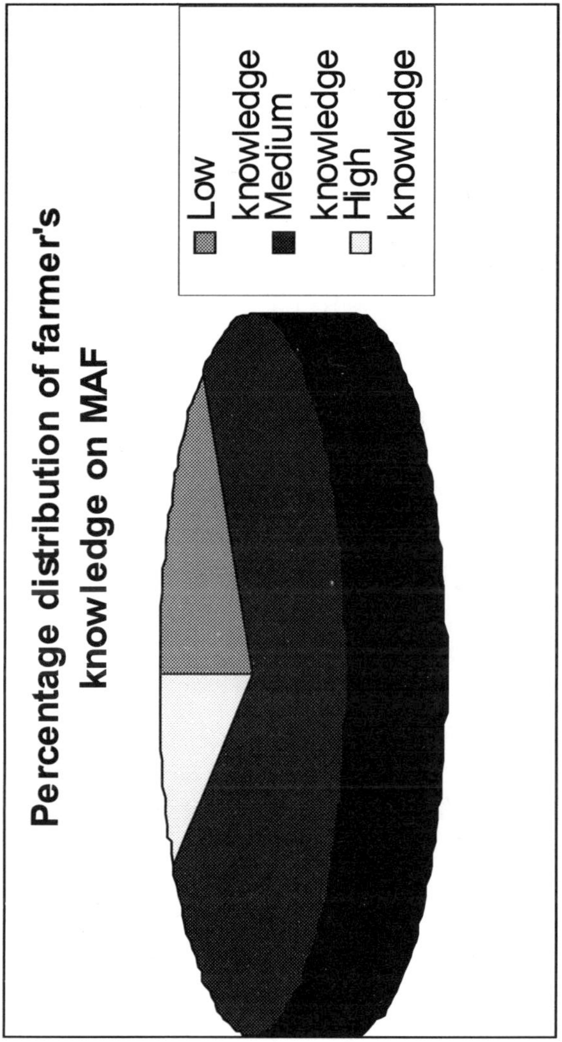

Figure 11.1: Pie Graph Showing Percentage Distribution of Farmers' Knowledge on Agroforestry Production System

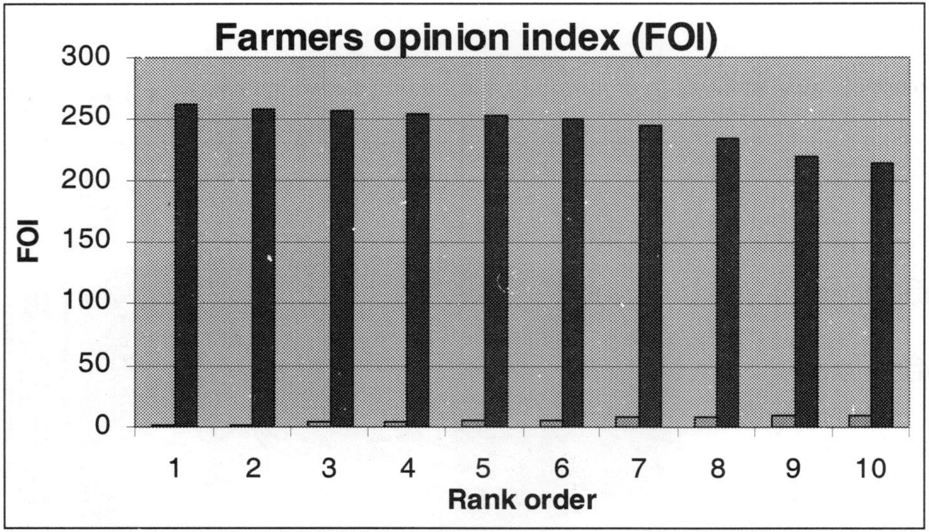

Legends: 1: Increase income; 2: Increase economic security; 3: Improve social status; 4: Improve local environment; 5: Increase the supply of timber and fodder; 6: Solve fuel problem; 7: Increase the availability of fruits; 8: Increase vegetable and spices production; 9: Increase employment opportunity and 10: Decrease family malnutrition

Figure 11.2: Bar Graph Showing Farmer's Opinion Index (FOI) on Socio-Economic Development through Agroforestry Production System

Expriment No. 2: The Role of Agroforestry System on the Improvement of Soil Properties in a Multilayered Coconut Orchard

Continuous 4 years practicing of agroforestry production system could not change soil pH remarkable over the initial stage at coconut garden. However, a slightly increasing trend in soil pH was observed under coconut based multistoried production system in the fourth season of under storey crop production. In general, higher soil pH was recorded in the tree-crop combinations of Coconut+Guava+Turmeric based multistoried production system (C+G+UC).

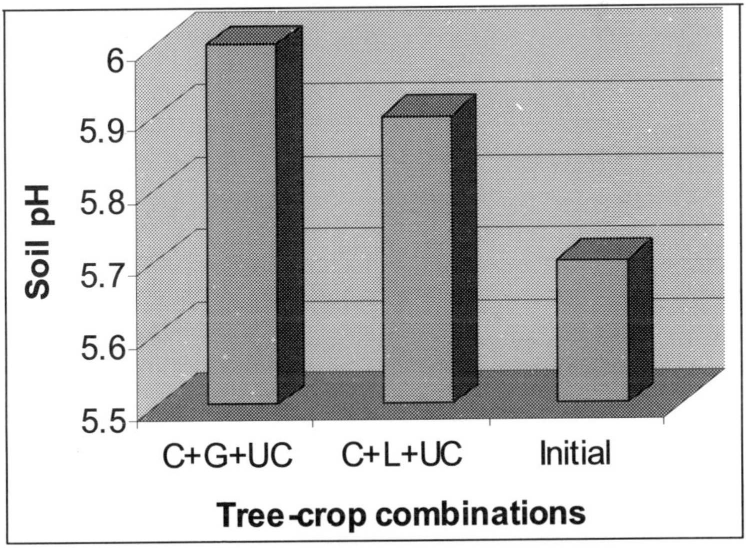

Figure 11.3: Soil pH Under Different Tree-Crop Combinations at Coconut Orchard

Soil organic carbon content followed also an increasing trend towards the multistoried agroforestry production system. It content under Coconut+Guava+Turmeric based multistoried agroforestry production system was found higher compared to the initial stage *i.e.* sole cropping of coconut. This may be due to differences in leaf and root litter shedding; rooting behavior and soil moisture content that influences decomposition of litters.

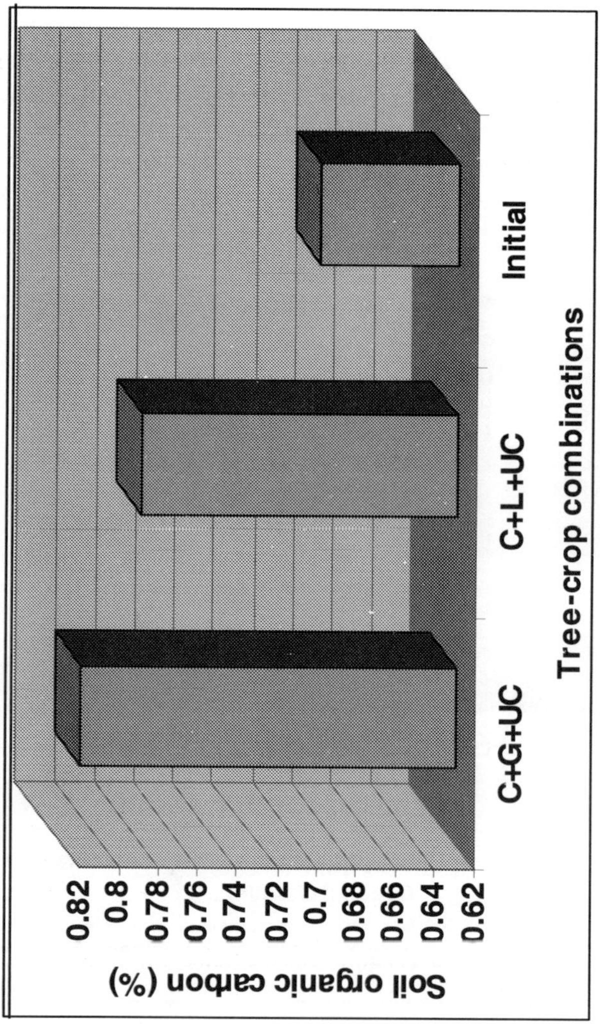

Figure 11.4: Soil Organic Carbon (%) Under Different Tree-Crop Combinations at Coconut Orchard

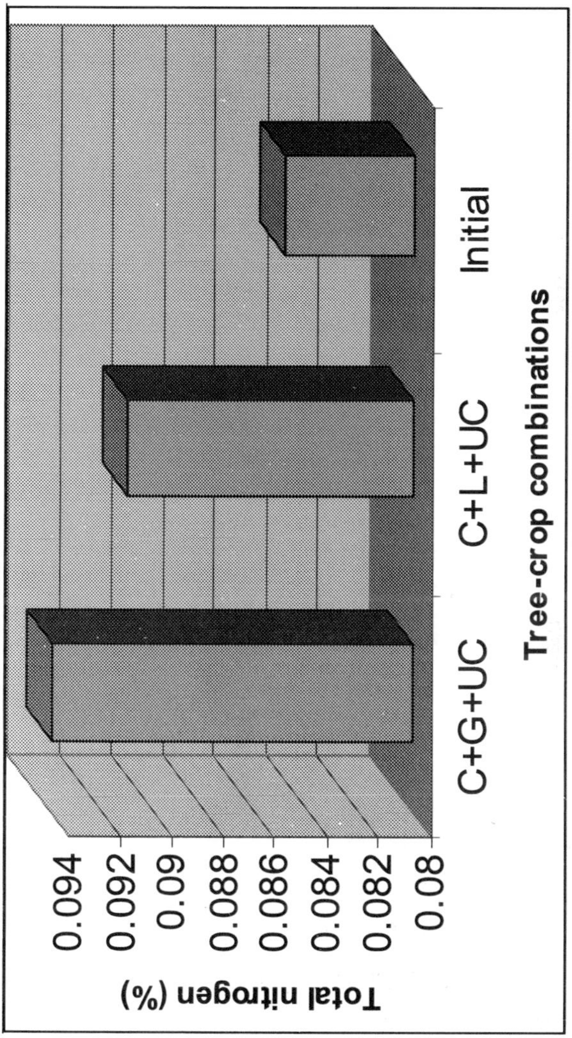

Figure 11.5: Total Nitrogen (%) Content in the Soils Under Different Tree-Crop Combinations at Coconut Orchard

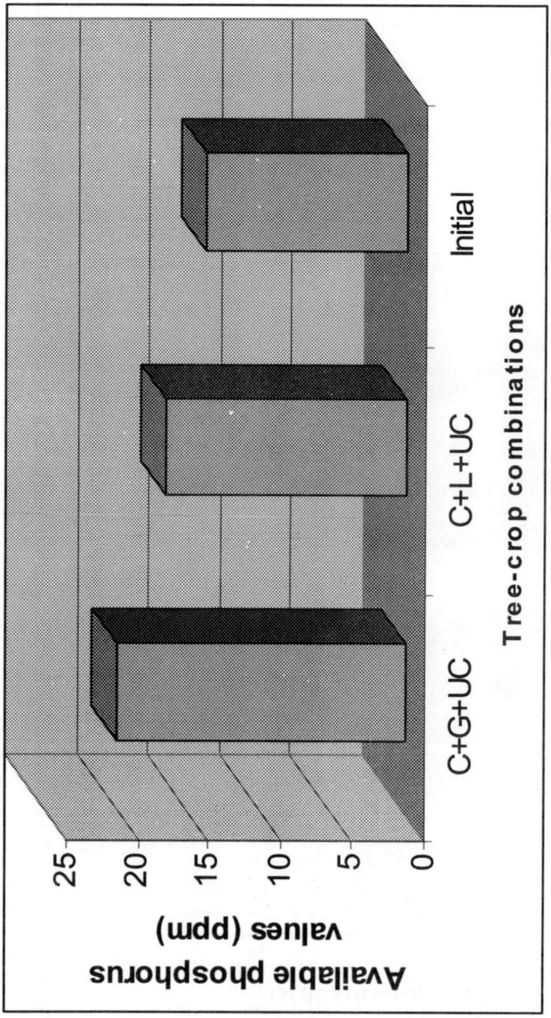

Figure 11.6: Available Phosphorus with Soil Under Different Tree-Crop Combinations at Coconut Orchard

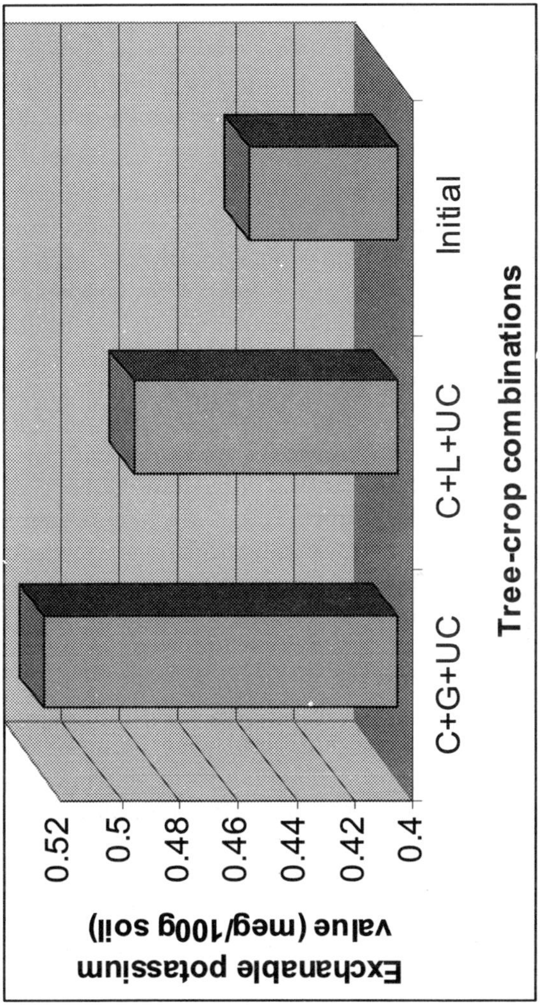

Figure 11.7: Potassium Content in the Soils Under Different Tree-Crop Combinations at Coconut Orchard

Total nitrogen content was slightly higher under agroforestry system. Such increasing trend occurred due to high moisture content and high LAI in agroforestry system. Amount of available phosphorus was also higher under the agroforesty production system in the fourth season of under storey crop production. Soil exchangeable potassium was 0.44 mg/100g soil in initial soil stage and it was found 0.52 mg/100g soil after the fourth harvest of understorey crops at Coconut+Guava+Turmeric based multistoried production system.

In case of soil chemical properties, the research study revealed that positive soil enrichment was occurred with respect to organic carbon, total nitrogen, available phosphorus and exchangeable potassium due to practicing agroforestry production system in coconut orchard. Coconut+Guava+Turmeric based multistoried production system had more positive effect on soil than that of Coconut+Lemon+Turmeric based multistoried agroforestry production system (C+L+UC). However, soil pH was somewhat slightly increased, whereas organic carbon has largely increased. Moreover, nitrogen, phosphorus and potassium also have followed the same trend with organic carbon. From the results, it was revealed that the changes of soil fertility are primarily similar to accumulation organic carbon. Possible processes of nutrient accumulation could be adsorptions in organic carbon components, or even natural accumulation processes on redeposit from leaf and litter fall as similar to the work of Willet (1995). While sources of organic carbon are explicit, from photosynthesis, sources of the nutrient could be recycled from depths through ground water system. Once the nutrients are released from the leaf litters, they could be reabsorbed by soil organic matter or leached to depth. In the latter cases deep rooting systems of trees in agroforestry system could reabsorb and conserve in the biomass before recycling back to the ecosystem. As such the agroforestry could improve the soil through intensive recycling and conservation of organic matter and nutrients.

Conclusion

From the findings of the present study, it may be concluded that the majority of rural farmers of the Barind tract of Bangladesh had a very good knowledge and medium extension contact related to agroforestry production system. It was also observed that 85 per cent farmers followed agroforestry production system and received benefits to a higher extent. Again, from the research study on orchard based mutistoried agroforestry cropping system, it was evident that there was an important effect of agroforestry cropping systems on the improvement of soil chemical properties. As a result, the agroforstry systems may be of help to attain the development objectives and preserve the biosphere core.

References

BBS. 2007. *Statistical Pocket Book of Bangladesh*. Bangladesh Bureau of Statistics, Ministry of Planning, Govt. of Peoples' Republic of Bangladesh.

Gillman, G.P. and Sumpter, E.A. 1986. Modification of the compulsive exchange method for measuring exchange characteristics of soils. *Australian Journal of Soil Research* 24: 61-66.

Maheswari S.K., Dhantonde B.N., Yadav S. and Gangrade S.K. 1985. Intercropping of Rauvolfia serpentina for higher monetary returns. *Indian J. Agri Sci.*, 58: 108-111.

Menzies, N. W, and Gillman, G.P. 1997. *Chemical Characterisation of Soils of a Tropical Humid Forest Zone: A Methodology*. Soil Science Society of America Journal 61, 1355-1363.

Pareek, S.K. and Gupta, R. 1984. Exploratory studies on yield and comparative economics of medicinal plants based cropping systems in north-western India. *Ann. Agric. Res.* 5: 169-177.

Sarkar, D.C. 1997. Correlates of Selected Characteristics of Potato Growers with their Adoption of Improved Potato Cultivation Prctices in five Villages of Comilla District. *M.S. (Ag. Ext. Ed.) Thesis*. Dep. Agric. Ext. Ed., Bangladesh Agric Univ. Mymensigngh, Bangladesh.

Willett, I.R. 1995. Role of organic matter in controlling chemical properties and fertility of sandy soils used for lowland rice in Northeast Thailand. In: *Soil Organic Matter Management for Sustainable Agriculture* (eds. R.D.B. Lefroy, G.J. Blair and E.T. Craswell) Australian Centre for International Agricultural Research Proceedings No 56, pp. 109-114. Brown Prior Anderson: Melbourne, Australia.

Chapter 12

Mysterious World of Microbial Diversity: Functioning for a Vibrant and Sustainable Eco-sphere

☆ *Poonam Gusain, Rajesh Kumar and Vir singh*

Biodiversity is undoubtedly the greatest wonder on Earth. It is not only an important resource but also strength of developing countries. It is an irreplaceable resource and if not conserved, its extinction is forever (Kalia and Gupta, 2005). The microbial world is the largest unexplored reservoir of biodiversity on the Earth. Microbes are essential for life since they perform numerous functions essential for the biosphere that include nutrient cycling and environmental detoxification. Yet the microbial life is the most mysterious and their mysteries are yet to be unfolded in their wholeness. The importance of microbial activities to the biosphere provides new avenues for understanding their diversity, conservation and exploitation (Satyanarayana, 2005). Convention on Biological Diversity (CBD), a treaty made in June 1992 Rio de Janerio Earth Summit, realized the importance of conservation of biodiversity giving special emphasis to microbial diversity: *"The beauty of bacteria is that they are always mutating and evolving into new and wondrous forms."*

Microbial Diversity: The Unseen Natural Resource

The biodiversity blossoming amongst plants and animals is visible and we often appreciate the same. Microbial diversity, which encompasses the spectrum of variability among all types of microorganisms involving bacteria, fungi, actinomycetes, microalgae, protozoans and other moneras, however is not visible by naked eyes. Neverthelss one gets non-plussed by an account of this unseen biodiversity if one gets exposed to it. Biodiversity, including that of microbes, blooms in the natural world, but human intervention results in substantial alteration. As the microorganisms are also the movers and shakers of global cycles of elements, they significantly contribute to stability and proper functioning of ecosystems that are threatened by man-made disturbances. Microorganisms exhibit extraordinary genetic and phenetic diversity. Seldom coming into picture and discussion, this biodiversity is an unseen natural resource that deserves greater attention. Despite their importance, less than 5 per cent of the world's microorganisms have been described.

The abundance of prokaryotes in open ocean, in soil and in oceanic and terrestrial sub-surfaces is 1.2×10^{29}, 2.6×10^{29}, $3.5\ 10^{30}$ and $0.25\text{-}2.5 \times 10^{30}$ respectively (Whitman *et al.*, 1998). Among the deep-sea micro flora approximately one third of the prokaryotes are Archaea (Karner *et al.*, 2001). The estimated number of viruses in the world oceans is approximately 7.5×10^{29} while no total virus number are known from soil and subsurface (Weinbauer and Hofle, 1998). The top 6 inches of fertile soil contains more than 2 tons of fungal and bacterial biomass. It has been estimated that there might be up to 10,000 bacterial species per 100 g soil (Torsvik *et al.*, 1990).

A handful of soil contains different kinds of microbes; even a single soil particle represents ecological niches for different types to develop (Steage and Zaft, 2002). Thus the prokaryotes have been described correctly as the unseen majority (Whitman *et al.*, 1998) and this is more correct for all microbes.

The Indian Biodiversity Scenario

The Indian subcontinent is one of the richest in terms of its biodiversity in the world owing to its vast geographic area, varied

Table 12.1: Beauty and Symmetry in Nature

Diversity	*Index*	*Examples*
Morphology	Shape	Rods, cocci, spiral, filaments squares, stars and most irregular forms
	Size	From less than 1μm to greater than 5μm
Structure	Prokaryote	Gram positive form, gram negative form, and archae form
	External features	Flagella, pili, stalks, capsules, sheaths
	Internal features	Spores, mesosomes, gas vesicles, crystals
Ecological	Habitat	Freshwater, marine, soil, deep subsurface
	Tolerance	Psychrophiles, mesophiles, thermophiles pH conditions
Behavioral	Motility and taxis symbiosis	Gliding, flagella
		Phototaxis, aerotaxis, chemotaxis
		Root nodules
Metabolic	Source of carbon and energy	Chemoautotrophs (carbon from CO_2 energy from light)
		Chemoheterotrophs (carbon and energy from organic molecules)
		Photoautotrophs (carbon from CO_2 molecules, energy from light)
		Photoheterotrophs (carbon from organic molecules energy from light)
Genetic		Similarity among genes, evolutionary linkages.

Source: http://www.mbio.ncsu.edu/JWB/MB409/home.html.

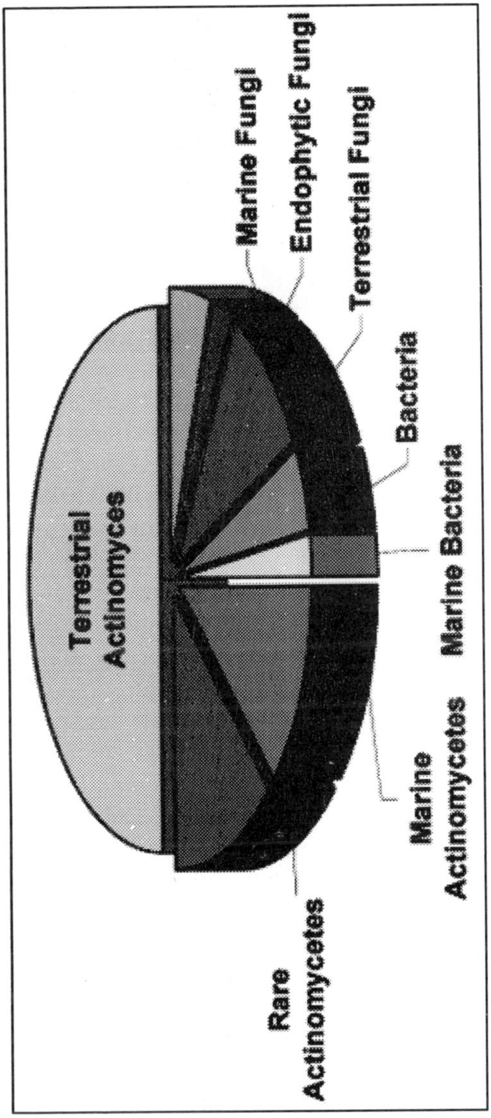

Figure 12.1: Distribution of Microbes

Source: http://www.microbialscreening.com/index.html

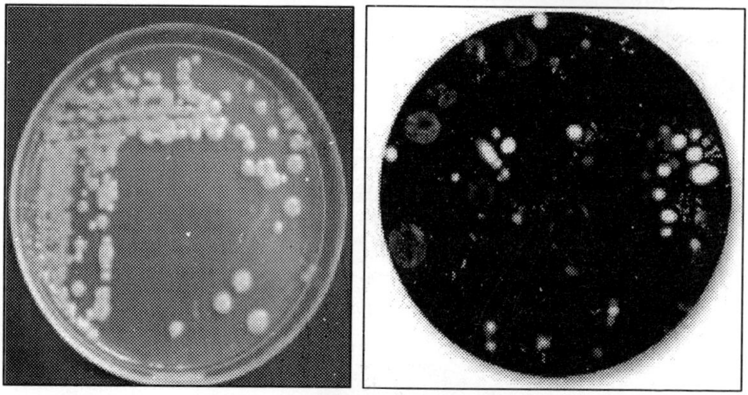

Figure 12.2: Bacterial Diversity

topography and climate. India has over 126,188 species of bacteria, fungi, plants and animals. Because of its richness in overall species diversity, India is recognized as one of the 12 mega diversity regions of the world. Nearly 72 per cent of India's biowealth is constituted by fungi (~18 per cent), insects (~40 per cent) and angiosperms (~13 per cent). Thus, India's contribution to the global diversity is around 8 per cent. The most important mega-diversity centers are Western ghats, North-eastern hill regions, Bastar regions inhabited by tribals, Andaman Nicobar islands, Mangrove forests of Sunderban area, Silent Valley of Kerala, Playas of Rajasthan, Chilka lake of Orissa, Sonar Lake of Maharashtra and the Himalayan region, are present in India that are home to diverse populations of microorganisms. It is interesting to note that extremely acidic soils (pH ~2.8) of Kerala are home to cyanobacteria. As many as 42 species were recorded in acidic soils of which 19 were recorded for the first time in Kerala ((http://www.nbaindia.org/docs/bulletin1-microdiversity.pdf).

Utility of Microbial Diversity

Microorganisms make a unique organization of life, like plants and animals. These organisms have been evolving for nearly 4 billion years and are capable of exploiting a vast range of energy sources and thriving in almost every habitat. It is estimated that 50 per cent of the living protoplasm on this planet is microbial. Microorganisms represent by far the richest repertoire of molecular and chemical diversity in nature. Microbial diversity existing in natural ecosystems has the following major applications (Kapur and Jain, 2004):

- ☆ They represent experimental systems with the strategies and extreme limits of life
- ☆ They are of critical importance in sustainability of life forms on earth
- ☆ Untapped microbial diversity is a resource for new genes and organisms of value to biotechnology
- ☆ Microbial diversity patterns of microbes can be used for monitoring and predicting environmental changes
- ☆ They help in conservation and restoration of the higher organisms
- ☆ Microbial communities are excellent models for understanding biological interactions and evolutionary history determines the stability of the biosphere, which in turn leads to stability in climate, water, soil, chemistry of air and overall health of the biosphere.
- ☆ Microbes are a source of species on which human race depends for food, fodder, fuel, fiber, shelter, medicine etc.
- ☆ Microbes carry out recycling of phosphorus, oxygen, carbon, nitrogen and sulphur elements and replenish the environment with these by degrading substrates obtained from dead and decaying plant and animal remains.

Conservation of Microbial Diversity in India

According to the World Conservation Monitoring Center, 1,604,000 species have been described at the global level. India accounts for 8 per cent of global biodiversity existing in only 2.4 per cent land area of the world (Khoshoo 1996; Varley and Scot, 1999).

The idea of establishing culture collections was conceived during the period 1880-1890. This is important from the point of view of the ready availability of basic microbial resource for carrying out fundamental studies, as well as for providing the raw material for developing novel products.

Conclusion

Despite the acknowledged value of microorganisms, our knowledge of their diversity and many of their key roles in sustaining global life support systems is still very scarce. There are isolated groups working on microbial diversity and microbial prospecting

Table 12.2: Enzymes Produced by Microbes for Industrial and Research Application

Enzyme	*Source/Origin*	*Microorganism*	*Reference*
Cellulase/Hemicellulase	With/without Tween-80	*Volvariella volvacea*	Cai *et al.* (1994)
Lignolytic enzymes	–	*Coriolusversi color*	Ding *et al.* (1994)
Laundry detergent (alkaline cellulose/protease)	Soil	Alkaliphile	Horikoshi (1999)
Cyclomaltodextrine gluanotranferase	Soil	Alkaliphile(cyclodextrine production)	Horikoshi (1999)
Extracellular lipase	–	*Pseudomonas aeruginosa* KKA-5	Sharon *et al.* (1999)
Keratinase	Feathers of birds	*Bacillus lichiniformis* PWD-1	Lin *et al.* (1997)
Alpha-galactosidase	S. cerevisiae	*Aspergillus awamori*, A. oryzae	Murphy and Power (2002)
Amylase	Culture filtrate	*Bacillus circulans* GRS-313	Dey *et al.* (2002)
Extracellular Deoxyribonuclease	NBG Media	*Streptomyces thermonitrificans*	Deshmukh and Shanker (1999)
Beta-xylosidase (Thermostable)	Birch xylan	*Thermomonospora curvata*	Stutzenberger F and Bodine (1998)
GTPcyclohydrolase II/3,4-DH-2-B-4-PS	–	*Bacillus subtilis*	Humbelin *et al.* (1999)
L-aspartate beta-decarboxylase	–	*Alcaligenes faecalis*	Chen *et al.* (2000)
Creatininase	*Pseudomonas putida* RS65	*E.coli*	Tang *et al.* (2000)

Contd...

Table 12.2–Contd...

Enzyme	*Source/Origin*	*Microorganism*	*Reference*
Proteinase	–	*Tritirachium album*	Fortelius and Markkenen (2000)
Serine hydroxyl Methyl transferase	–	*E.coli*	Kreuzman *et al.* (1997)

Compiled by different sources by Kalia, and Gupta (2005)

* References quoted in the source not listed in this paper.

all over the country. But there is no systematic effort to take up major projects on these aspects. Exploration, evaluation and exploitation of microbial diversity is essential for scientific, industrial, ecological and social development.

Table 12.3: List of Culture Collection Centers in India

CCDMBI	Culture collection, Department of Microbiology, Bose Institute, Kolkata
CIPDE	Collection of Insect Pathogen, Department of Entomology, Marathwada Agriculture University Parbhani
BDC	Biodiversity Documentation Centre, Jawaherlal Nehru Centre for Advances Scientific Research Jakkur, Bangalore
CCMB	Centre for Cellular and Molecular Biology, Hyderabad
DRDO	Defense Material and Stores Research and Development Establishment Culture Collection, Defense Research and Development Organization, New Delhi
IISC	Indian Institute of Science, Bangalore
DBV	Division of Standardization, Indian Veterinary Research Institute Izatnagar
DMSRDE	Defense Material and Stores Research and Development Establishment, Kanpur
DUM	Delhi University Mycological Herbarium, Delhi
ITCC	Indian Type Culture Collection, Indian Agriculture Research Institute, Delhi
MCM	Macs Collection of Microorganisms, ARI Pune
MTCC	Microbial Type Culture Collection and Gene Bank, IMTECH Chandigarh
NIBAIM	National Bureau of Agriculturally Important Microorganisms Kusmaur
NCDC	National Collection of Dairy Cultures, National Dairy Research Institute, Karnal
NCIM	National Collection of Industrial Microorganisms, National Chemical Laboratory Pune
RRL	Regional Research Laboratory, Jammu
UMMFTD	Food and Fermentation Technology Division, University of Mumbai
VPCI	Fugal Culture Collection, Vallabh Bhai Patel Chest Institute, Delhi

Compiled by Lal (2007).

Revealing Some Mysteries about Microorganisms

- ☆ Typically there are between 10,000 and 10 million bacteria on each hand.

- ☆ Bacteria 40 million years old have been extracted and successfully grown from a fossilised bee.
- ☆ In 1918 more people died from the influenza virus (approximately 30 million) than died in the First World War (10 million).
- ☆ Almost one million bacteria can be created by one person in a school day.
- ☆ Nearly a third of all bottled drinking water purchased in the US is contaminated with bacteria.
- ☆ If you urinate when swimming in a South American river, you may encounter the candiru. Drawn to warmth, this tiny fish is known to follow a stream of urine to its source, swim inside the body, and flare its barbed fins. It will remain firmly embedded in the flesh until surgically removed.
- ☆ The soft plastic headphones used on airplanes create a warm, moist environment in the ear canal that is ideal for breeding bacteria. Wearing headphones for just an hour will increase the bacteria in your ear by 700 times.

Acknowledgement

We earnestly appreciate the mysterious microbes that rule over the world's ecosystems. Humanity is non-plussed by the unique role of microbes. No matter where we go or what we're doing, we cannot escape them. Without the microbes life, as we know, would cease to exist.

References

Finlay, B.J. and Esteban G.F. 2001. Ubiquitous microbes and ecosystem function. *Limnetica* 20(1):31-43.

http://www.nbaindia.org/docs/bulletin1-microdiversity.pdf

http://www.mbio.ncsu.edu/JWB/MB409/home.html

http://www.microbialscreening.com/index.html

Kapur, M. and Jain, R. 2004. Microbial diversity exploring the unexplored. *Imtec Communication*, No. 08.

Kalia, A. and Gupta, R.P. 2005. Conservation and utilization of microbial diversity. *NBA Scientific Bulletin*, No.1: 1-40.

Khoshoo, T.N. 1996. India needs a National Biodiversity Conservation Board. *Curr Sci*, 71 (7):506-513.

Lal, R. 2007. Microbial diversity, systematic research and culture collection facility in India. *Indian Journal of Microbiology*. 47:i-ii.

Markus, B. Karner, Edward F., Delong and Devid, M Karl. 2001. Archael dominance in the mesopelagic zone of the Pacific ocean. *Nature*, 409: 507-510.

Satyanarayana, T. 2005. Meeting report on microbial diversity. *Current Science*, 89(6):926-928.

Steage, H.T. and Zaft, R. 2002. Density and diversity. *Nature* 417: 698-699.

Torsvik, V., Goksoyr, J. and Daae, F.L. 1990. High diversity in DNA of soil bacteria. *Appl Environ Microbiol*, 56: 782-787.

Varley, J.D. and Scott, P.T. 1999. Conservation of microbial diversity: A Yellowstone priority. *ASM News* 64(3): 147-151.

Whitman, W., Coleman, D and Wiebe, W. 1998. Prokaryotes: The unseen majority. *Proc Natl Aca Sci USA*, 95: 6578-6583.

Weinbauer, M.G. and Hofle, M.G. 1998. Distribution and life strategies of two bacterial populations in a eutrophic lake. *Appl Environ Microbiol*, 64: 3776-3783.

Chapter 13

Biodiversity Conservation through ICT with ITK Use

☆ *Sucheta Arora and Asheesh Chauhan*

India has suffered a huge loss due to neglect of traditional knowledge system and degradation of biodiversity. Technology is not only a mechanical innovation but also an indigenous skills and metabolic innovations which mostly come from local people. To prevent the deterioration of natural resources and environment and to promote their sustainable use, documenting appropriate local technologies and indigenous traditional knowledge is essential. The ICT can be used to create and increase awareness and understanding of biodiversity and ITK to motivate people in this direction. Resource rich and conscious people may get information through Internet. To build capacity of various stakeholders for sustainable development, create expertise in the field of environmental education and in developing innovative participatory learning methodologies, ICT can be used. Online services can also be provided through IT for information, education, training monitoring, consultation, and diagnosis. ICTs can be helpful in providing the interaction among local communities, policy makers and scientists. Since the IT has wider applications in science and technology, it will be used for scientific validation of ITKs.

India harbours unique ecosystems. Different parts of the county are storehouse of gene pools. Many varieties of plants and animals have originated in India. People here cultivate many lesser known crops and plants. This biodiversity and indigenous knowledge is of immense value for future. Despite the richness of biodiversity and indigenous knowledge no proper strategy has been formulated to conserve biodiversity and promote indigenous technical knowledge (ITK). Even in mountain regions, where biodiversity is quite rich, the per capita income is the lowest in the world because of ignorance.

Traditionally, people achieved sustainable lifestyles by strategically rotating diversity, both physical and biological, to counter environmental risks and uncertainties. Environmental resources played an important safety function against environment stress and exogenous shocks.

Indigenous knowledge is the term often used today to designate a form of knowledge, which is holistic and orally transmitted within local communities. It is also called 'traditional' and 'local knowledge' to differentiate this knowledge from the international knowledge system also called the 'western' or scientific 'system that has been generated by universities, government research centers and private industries.' Indigenous knowledge system refers to the ways in which members of a given community define and classify phenomena in the physical, natural and social environment. Such knowledge provides the background for local level decision-making. Traditional knowledge refers to unique local knowledge, values and technical capabilities existing within and developed around the specific conditions of communities, indigenous to particular around the specific conditions of communities indigenous to particular area and has existed over centuries but largely remains confined to our home and village. It is only in last two decades that there has been a great awakening to recognize and include traditional knowledge in formal stream of knowledge basket.

India has suffered a huge loss due to neglect of traditional knowledge system and degradation of biodiversity. Indigenous knowledge and rich biodiversity are the pearls of wisdom.

Challenges and Action Needed

Adaptability and suitability should be the primary concern of technologies selected for implementation. However, in most developmental programmes, ideas and technologies developed in one area/region and transported to another region there is no regard of their adaptability and suitability. The technologies are not pro-people and pro-poor. Technology is not only a mechanical innovation but also an indigenous skills and metabolic innovations which mostly come from local people. Thus there is need to identify local innovations both technologies as well as institutional structures, which can be mainstreamed in diverse agro-climatic zone across the country. In hilly areas, mountain is harsh and always challenges its inhabitants. They have to respond it creatively and with ingenuity for their survival. In this process mountain communities have discovered unique ways and means and natural resources management practices to survive in these areas. Their indigenous knowledge is a rich heritage with inputs from modern sciences and technology. They well provide us clue to manage our mountain areas better and for having equitable and sustainable development. The kind of interest is scorned up on by most others because as yet there is no international legal framework to guarantee the intellectual property right of local people who are approached by outsiders for the shake of profit. But there are also few attempts by national governments to record any respect to this knowledge of wild plant and plant genetic resources. Few address indigenous knowledge as something which is the part of culture and which therefore could be strengthened so that people can maintain their cultural integrity including belief system. For instance 'People Biodiversity Register Programme' is a programme of organizing information in the domain of folk science.

Another issue associated with ITK is poor identification and documentation. Without proper identification and documentation, their use and promotion will be out of question. Thus, location specific projects on memory harvesting should be taken up for identification and documentation of traditional knowledge. Scientific validation of such knowledge is also required.

Natural resources need be prevented from their deterioration which is a must for promoting their sustainable use and documenting appropriate local technologies. In the same line 'People Biodiversity

Register Programme' was focused on creation of decentralized countrywide database on status of biodiversity resources, on-going involvement of local communities and individuals in sustainable use and conservation of biodiversity, local knowledge, etc. In the long run this programme is expected to validate the information recorded; for instance through cross checking with other published information sources, create a network of decentralized database, ultimately linking them to a consolidated national database which would be given full credit to the origin of information, feedback pertinent information back to local communities and contribute to country wide monitoring of national biodiversity resources.

Absence of database on biodiversity and networking among different partners associated with biodiversity is also a big challenge. Thus now it becomes essential to prepare a database on biodiversity resources and ITKs. Such database will be of no use unless shared with people, policy makers and other partners including NGOs involved in biodiversity and ITK. Thus, the need of sharing calls for proper networking.

Exploiting Potential of ICT

Information technology is a mix of computer and communication technologies and is mainly used for processing, storing and retrieving of information in multimedia format. The IT revolution has the potential to ensure that knowledge, information, technologies, methods and practices are put in right hands. Taking advantage of improved access and communication technologies, people can greatly benefit from establishing links between themselves and local institutions.

The rapid growth of mass media and coverage of radio and TV also offer a major opportunity for environmental education. The increasing availability of Internet is also an opportunity with the media, giving more coverage to environmental issues.

Awareness about the conservation of biodiversity is an urgent need of our times. The ICT can be used to create and increase awareness and understanding of biodiversity and ITK to motivate people in this direction. People may be sensitized through video show and other media. Resource rich and conscious people may get information through Internet. To build capacity of various stakeholders for sustainable development, create expertise in the field of environmental education and in developing innovative participatory learning methodologies, ICT can be used. This will create a cadre of people who can work for the cause of biodiversity conservation and ITK promotion.

The ICT can be helpful in identifying and popularizing appropriate technologies and practices for ensuring sustainable livelihood. For documentation of information on traditional knowledge and disseminate it through organized networking is also required. The ICT has the potential to undertake these effectively. Mapping of biodiversity resources and ITK and preparation and maintenance of such database may be of immense use. This database needs to be updated continuously. With the advancement in ICTs such mapping and database preparation is possible.

The effective training and education in the field of environmental education can also be imparted effectively with the use of ICTs. This will strengthen the mechanism to impart the basic skills needed for biodiversity conservation and ITK promotion for a better future. Online services can also be provided through IT for information, education, training, monitoring, consultation, and diagnosis. ICTs can be helpful in providing the interaction among local communities, policy makers and scientists. Tele-education through IT is also possible. Since the IT has potential to localize the global content and globalize the local content, the exchange of information and sharing of experiences will be an easy task. Since the IT has wider applications in science and technology, it will be used for scientific validation of ITKs.

For all this communication material on biodiversity and ITK needs to be developed for creating awareness and interest among different target groups. The concept of tale-centres may also be used for these purposes. There is evidence that tele-centers have played a major role in mobilizing communities to address their developmental problems. These centers can be used as information hubs that capture, repackage and disseminate information to rural communities. Strategic use of available resources including ICTs should be promoted to conserve biodiversity and promotion of ITKs for sustainable development.

References

paradiso-fp7.eu/wp-content/plugins/alcyonis-event.../Jeff_McNeely.pdf

www.chriscoldewey.com/tag/ict-for-conservation/

www.t4cd.org › t4cd - Home › Resources › ICT Resources

www.without-borders.org/?q=node/708

ict-ensure.tugraz.at/en/index.php/ensure/Content2/.../Biodiversity.

Chapter 14

Biodiversity and Traditions at the Heart of Sustainability: Farmers' Way of Life in the Himalayan Mountains

☆ *Vir Singh*

Introduction

A wrong attitude to nature
Implies somewhere a wrong attitude to God
And the consequence is an inevitable doom.

– *T.S. Eliot*

Himalayas are the youngest, the highest, the most fragile, and the most vulnerable mountain ecosystems on Earth. These ecosystems are inhabited overwhelmingly by farming communities throughout their length and breadth. Agriculture and pastoralism are the major occupations and sources of livelihoods of the people.

Socio-economic, cultural and ecological aspects of mountain life are altogether distinguishable from those of the mainstream plains. Marginality, fragility, limited accessibility, and diversity or heterogeneity are the key characteristics of these ecosystems. These characteristics offer opportunities, but at the same time also impose certain restrictions and limitations on the production and livelihood systems.

One of the major, perhaps outstanding, characteristics that mountain areas embrace is the human adaptation mechanisms (Jodha *et al.*, 1992; Singh and Tulachan 2001). This helps farmers cope with the limitations imposed by marginality, fragility and high degree of inaccessibility that mountain area face.

Mountain areas have been in the throes of vicious cycle of ecological and socioeconomic crises. One of the major responses to address the socioeconomic problems was the adoption of the Green Revolution technologies beginning since mid-1960s. But, this was no panacea for the farming communities in the mountains. Instead this contributed to several other ecological, environmental, and socioeconomic problems in the fragile ecosystems.

Mountain farmers face the multiple crises arising out of geopolitical, biophysical, and socioeconomic marginality (Singh and Tulachan 2001). The marginal farmers have their own ways and means to combat and reverse these crises. This was demonstrated by masses in Indian Central Himalayas (recently carved out as a separate Indian Himalayan State of Uttaranchal) in the form of the oft-quoted Chipko Andolan (Hug the Trees Movement) (Bahuguna 1995). Average mountain farmer is a natural eco-philosopher and tends to be innovative in his/ her approaches to life and livelihoods. Conservation and sustainable use of biodiversity in the land-based livelihood systems are central to local farmers' way to living. Traditional strategies marginal mountain farmers have evolved over millennia are sustainability-inducing. The traditional strategies need to be preserved, documented, demystified, recognised and promoted. This paper examines some of such strategies. The traditional strategies need not be regarded as a remnant of the past, but something that link us with blossoming future. They enlighten us to a path towards transcendental development and sustainability.

Farmers' Perception of the Green Revolution

The direction of science and technology has been to find new ways and means to exploit nature. This has resulted in the depletion of resources, both renewable and non-renewable, due to overexploitation to satisfy the insatiable demands of society. This development created temporary prosperity, but has also created pollution, war and poverty.

– Bahuguna (1995)

The conventional agriculture technology – the Green Revolution technology – was based on the assumption that technology is a superior substitute for nature, and hence a means of producing limitless growth, unconstrained by nature's limits. However, the assumption of nature as a source of scarcity, and technology as a source of abundance, leads to the creation of technologies which create new scarcities in nature through ecological destruction. The reduction in availability of fertile land and genetic diversity of crops as a result of the Green Revolution practices indicates that at the ecological level, the Green Revolution produced scarcity, not abundance (Shiva 1991, Shiva *et al.*, 2005).

Mountain areas in the Himalayas, thanks to their terrain conditions and high degree of inaccessibility, did not permit Green Revolution to rein throughout. It has been possible only in the valleys equipped with irrigation facilities. These favourable valleys have witnessed almost complete genetic transformation of food crops, especially of rice and wheat. Some market-linked valley areas have been transformed into vegetable belts. Some of the patronised high-altitude areas have been converted into fruit belts with apple as the only fruit species. Only the mid-altitude rainfed areas have continued to be traditional.

The agricultural transformation in the mountains was officially regarded as a panacea for food security in the food-deficient region. Replacing the traditional crops by market-oriented cash crops was, however, could neither fetch food self-sufficiency to the region, nor could increase purchase power of the people. Mountain farmers' experiences of Green Revolution have been quite bitter. Having adopted the Green Revolution in favourable areas, they have only harvested crops of hunger, malnutrition, and discontent. Some of the things they would recount as their experience about the Green Revolution are as follows:

- change in (agri)cultural values;
- narrow straw-grain ratios leading to fodder shortage to livestock and drudgery on women;
- high degree of vulnerability for crop damages due to high incidence of epidemics and unfavourable weather conditions;
- pollution of the environment leading to health hazards;
- knocking down of soil flora and fauna and increasing incidence of desertification;
- poor nutritive value and less palatability of new food grains;
- ceasing of seeds to be common property resources;
- no free exchange of germplasm within community;
- decrease in collective responsibility;
- breakdown of social sanctions on resource usage;
- slackness in social cohesion;
- dependence on expensive, often unaffordable, external inputs and technology; and
- high degree of risks.

Having experienced unsustainability and food insecurity associated with the monoculture-based Green Revolution, the mountain farmers are now opting for biodiversity-based farming practices. Owing to the poor accessibility and other terrain features of the Himalayan Region, many remote, isolated, and somewhat inaccessible mountain areas did not witness a genetic wipe out as the transformed ones did. Mountain farmers for many years have been collecting their lost seeds from these areas and re-introducing them in the transformed areas from where these had vanished thanks to the introduction of fertilizer-responsive high yielding varieties (HYVs). Deprived of their genetic wealth, many Green Revolution areas in the mountains of Uttaranchal are once again boasting of their rich agrobiodiversity (Singh 1998; Singh and Jardhari 2000).

CPR-Centred Farming

Forest areas in the mountains are also a source of high quality food. These uncultivated areas, in fact, have potential to provide more quality food per unit area on sustained basis than the cultivated land. People in the mountains have been obtaining food (edible fruits, flowers, buds, seeds, vegetables, mushrooms, honey, etc.) from the forests for centuries. For the management of uncultivated areas for food, CPRs should be in our focus.

– Singh (2001)

Marginal mountain farmers throughout the Himalayan Region possess the area of arable land that is much less than that could provide them food security and a stable livelihood base (Partap 1999; Singh and Tulachan 2001). One of the essential features of mountain farming systems is the large uncultivated area, also known as the support area, rendering enormous ecological and economic services to the Region. The ratio between uncultivated and cultivated area is wide enough, further signifying the existing and prospective role of the former compared to that of the latter. Vast areas under the uncultivated land are the common property resources (CPRs). Local farmers have access to and some defined rights on these commons.

Mountain farmers have evolved over centuries the farming systems and the strategies that are compatible with the specific resource features of the mountain ecosystems (Jodha *et al.*, 1992; Singh 1998). CPRs – mainly the community lands under forests and grasslands – have been central to these strategies. Marginal farmers in the HKH Region are amongst the poorest communities of the world. They are, however, not resource-poor per se. They have access to and defined rights on common property resources (CPRs) under uncultivated support areas. Mountain farmers, from this yardstick, can be regarded as resource-rich. Their current poverty and pitiable condition are mainly attributable to their resource base degradation and continued negligence of the CPRs.

CPRs are at the heart of mountain agriculture. CPR-farmland linkages infuse vitality in the whole system. The uncultivated CPR lands provide a natural subsidy to the farmlands via the agency of livestock. These provide fodder to livestock due to which no proportion of cultivated area in the mountains has to be devoted to fodder cultivation. In other words, there is no man-livestock competition on marginal farms. Only what is non-consumable products for human beings (crop residues, weeds, etc.) to grow in the farmland are provided to livestock. Almost entire fuel wood is extracted from the CPR areas, in the absence of which livestock dung would be used as a fuel for cooking to the impoverishment of soil fertility.

Ecological niches full of exceptionally high-value biodiversity – including CPRs, and genetic resources in the farmlands – would form the basis for the sustainable growth of mountains' own economies. Mountains' biodiversity, indeed, can also contribute to national, regional, and global economies. Marginality- and inaccessibility-ridden mountain farming systems will remain one of the richest repositories of biodiversity on Earth in future and thus will be in a state to extend services to humanity in the foreland mainstream areas that are losing production potential due to serious erosive effects of monocultures of the Green Revolution combined with non-contribution/ absence of forests/ CPRs.

CPR-Farm Linkages

Agro-ecosystems serve as the strongest foundation on which to design strategies for transcendental and sustainable agriculture in mountain areas.

*– **Singh (2005)***

The biophysical resource base – land, soil, water sources, forest ecosystems, and biodiversity – is basic to the sustainability of the food production and livelihoods in the mountains. Forest-cropland ratio must be as wide enough as possible (dense forest area, in order to provide a sound ecological balance and natural subsidy to farmland, must be at least five times larger than the cropland area). Forest ecosystems provide stability and give vitality to the whole food production system by maintaining strong organic linkages amongst livestock, cropland, and human beings. Living soil, one of the basic capitals for sustainable agriculture, is the product of forest ecosystems. Continuous biomass flow from the forest areas (bedding material in animal sheds and fodder converted into manure, for example) enriches the soil ecosystem, a process that supports biodiversity and manifests into crop production (Tulachan and Neupane 1999; Singh 2000). Forests play a crucial role in nature's hydrological cycle and thus water sources in the mountains are also largely a product of the forests. Flow of water and moisture circulation within a farming system is vital for the sustenance of all forms of life. Availability of water in the system also ensures higher level of food production.

This diversity of the farming systems that exists in the mountains to this day must be preserved and enhanced. An appropriate forestland-cropland ratio exists in most of the mountain areas in the Himalayan Region. Forestlands, however, are mostly ecologically degraded, witnessing only poor canopy cover or no vegetation cover at all, on large areas. Any expansion of the area of cropland at the expense of forestlands would only disturb the delicate ecological balance. Actually, an increase in the cropland area becomes necessary when the production approach is commodity-centered with only cropland isolated from CPRs in focus, as in the case of the Green Revolution-type agriculture in the plains. Improvement in the condition of the forests in the Region, however, would result into substantial increase in productivity of the croplands through increased flow of fodder biomass and water/ moisture.

Forest areas in the mountains are also a source of high quality food. These uncultivated areas, in fact, have potential to provide more quality food per unit area on sustained basis than the cultivated land (Singh 1991, 2005). People in the mountains have been obtaining food (edible fruits, flowers, buds, seeds, vegetables, mushrooms, honey, etc.) from the forests for centuries. For the management of uncultivated areas for food, CPRs should be in our focus. As evident from the perusal of Table 14.1, CPRs are the richer source of food than the cultivated land. And, after all, obtaining food from the uncultivated land/ CPRs means to depend on a more stable system than from the more vulnerable cultivated land.

CPRs thus offer to provide the largest number of edible food products (189). Numbers of plant species being cultivated in the cropland/ cultivated area (92) in mountain agro-ecosystems are also more that the ones grown in plain agriculture (33). Mountain agro-ecosystems in general and CPRs in particular, thus, are especially rich in the biodiversity of specific consumptive use values.

Diversity-complexity is the foundation of agricultural sustainability. Modernisation of agriculture has weakened this foundation of Himalayan agriculture. This has changed the state of diversity-complexity into monoculture-simplicity, and sustainability into unsustainability. Plains' agriculture, generally devoid of the CPR component, is less diverse and simpler than that of mountain agriculture. CPRs with natural green mantle provide ecological stability to the agriculture. Its missing in the farming system, as in the transformed plain areas, makes the whole system vulnerable to damage.

Table 14.1: Number of Plant Species Providing Food in the Mountain Ecosystems and the Plains' Agriculture

Food Item	*Mountain Ecosystems**		*Plains' Agriculture***
	CPRs	*Cropland*	
Fruits	24	18	6
Vegetables	14	32	10
Cereals, Millets and Pseudo cereals	–	12	4
Pulses	–	10	5
Oilseeds	–	7	4
Medicinal and Aromatic Plants	99	4	–
Fodder***	52	9	–
Total Species	189	92	33

Source: Singh (2001).

* Figures are based on a survey of Henwal Valley in the Garhwal area of the Uttaranchal Mountains.

** Figures are based on a survey of the plains adjoining the mountains of Garhwal.

*** Livestock convert this forest biomass into human food, *e.g.*, milk.

Productivity of fruit trees that form dominant phytocommunity is also many times more than that of cereal crops. For instance, when the land is used to produce animal protein, an acre of land can yield a hundred kilograms of beef in one year. The same land will produce one to one-and-a-half tones of cereals, seven tones of fruits or 10 to 15 tonnes of walnuts. If trees are grown to give edible seeds for human beings and livestock, the same land can grow 15 to 20 tonnes, besides enhancing the fertility of the soil (Bahuguna 1987-88).

Plantation of food-yielding trees, shrubs and herbs in the CPR areas would enhance biodiversity and the ecological stability of the farming system along with the food security of the marginal farmers depending on them. Food products obtainable from trees are generally more palatable and of higher nutritive value than those provided by cultivated crops. Moreover, unlike the foodgrains and vegetables to come from the cultivated land, the variety of foods occurring in the CPRs does not require cooking or any other energy-consuming processing. Cultivated foods, contrarily, require fuel wood extraction from the forest areas, which often is one of the major causes of the state of ecological imbalance in the area. Food obtainable from forest area, however, does not make any such demand. Furthermore, the management of uncultivated areas for food by introducing food-providing woody perennials is ecologically a more stable system of food production than mere the cultivation of cereal crops on cultivated land.

Using the Biodiversity of Resource Base

Responsibility for life at large obliges us to support diversity, for diversity is not only the spice of life, it is also a precondition of vibrant and healthy life; moreover, a precondition of the meaningfulness and richness of our individual lives.

*– **Skolimowski (1995)***

Agroecosystems with many different niches occupied by many different kinds of species are more stable than those with only one species (as in monocropping) and thus give the farmers more security. This is the principle marginal farmers stick to minimise risks and increase the level of their food security. They always tend to use extremely large number of useful plant species that can thrive in a particular farming situation. They also maintain functional diversity, achievable through combining complementary plant and animal species in synergetic interactions, which is one of the means to inject sustainability into the agroecosystems. Each plant species has very broad base of genetic diversity.

"Diversity is prosperity" is a traditional adage in the mountains. *Baranaaja* culture raised by marginality-ridden mountain farmers is a unique testimony of it. *Baranaaja* literally means twelve foodgrains intercropped with finger millet as the base crop. Amaranth, buckwheat, kidney beans, horse gram, black soybean, black gram, green gram, cowpea, rice bean, adzuki bean, sorghum, and *cleome are* the main crops intermixed with finger millet. In some areas even groundnuts are also mixed with the base crop. *Baranaaja* provides a unique example of how a mountain farmer cultivates biodiversity. This management helps provide the maximum number of food items ensuring balanced diet from the minimum area of land. This also represents one of the key points of the strategies involving nutrient management in the system that cares human nutrition as well as health of the ecosystem.

Diversity-laden agriculture is the best bet for reducing risks and enhancing the degree of security. The gains accrued to the marginal farmers through diversified agriculture, however, are limited by the size of the arable land they own. But, it is not the arable land alone that could serve as the base for food security as is perceived in the context of the plains. Marginal farmers in the mountains give equal, rather greater, importance to the CPRs, other marginal ecosystems (alpine meadows, for example), and water sources.

CPRs are often appreciated for their contribution to the community of fuel, fodder, and timber. That they can provide the most basic necessity of humans, i.e., a variety of food items, is seldom appreciated. Vital role of uncultivated forest areas for providing different food items to the community is evident from the example of the Henwal Valley of Garhwal in Uttaranchal Mountains. In this Valley as many as 189 different naturally occurring species of plants offer edible products while people cultivate only 92 different food-producing plants on their cultivated land. Comparing this scenario with that of the plains' agriculture in the adjoining areas of the Garhwal Mountains, we would find that farmers depend only on 33 plant species for their survival (Table 14.1). Thus, from the point of view of plant species' diversity offering food articles, CPRs are the richest sources.

Each food crop species has a very broad genetic base. Marginal farmers cultivate an appreciably large number of varieties or landraces of every crop. Himalayan valleys of Uttaranchal are especially famous for a variety of rice landraces, each with specific traits. Traditional rice varieties, like *Hansraj, Ramjawan, Kanguri, Bangway, Gorakhpuri, Basmati, Thapachini, Jhumkya,* etc., thrive in lowland valley areas, while *Chawaria, Munjil, Jhailda, Lekmal, Kallao,*

Almunji, Chwatu, etc., grow well in upland rainfed areas and at high altitudes. Some landraces can be grown even close to glaciers. Some landraces demand more water, some less, and some need no irrigation at all. Productivity of rainfed varieties is comparable with those of irrigated ones. Such rare, hardy, and sturdy varieties would hardly be found in the plain areas anywhere in the world (Singh and Jardhari 2000).

Marginal farmers rear all types of animals – draught animals, carry-pack animals, milch animals, wool animals, meat animals, poultry, etc. – that are suitable for the farming systems. Functional diversity, mountain farmers achieve through combining complementary plant and animal species in synergetic interactions, injects sustainability into the agroecosystems (Singh 1998b). Livestock form the core of livelihood systems of livestock-dependent marginal mountain communities. They are not just an economic source, but are farmers' best companion serving them as manurers, exploiters of wastes, sources of power, forms of investments, and in many more ways. They form a cultural identity of mountain people. By transferring nutrients from forest ecosystem to farmlands and maintaining cyclic flow pattern in the farming system, they contribute to the ecological integrity of the system (Singh 1998, 2000; Tulachan 1998; Tulachan and Neupane 1999).

High degree of biodiversity creates barriers against any natural calamity (drought, for example) and pest epidemics. Some living organisms may be host to certain insects or diseases, some may be little more prone to weather extremes or other adverse conditions, but a complex mix of naturally occurring (in natural forest areas) organisms will do away any chances of widespread loss to the farmers. Marginal farmers in the mountaineous State of Uttaranchal in India would cultivate many-many different varieties for different purposes (productivity, drought resistance, taste, colour, nutritive value, rituals, etc.) and in doing so they enhance the degree of their security.

Creating Complexity or Intensive Use of Farm Elements

Western modes of economic development are not benign scientific models but hidden vehicles of violence.

– Skolimowski (1997)

Complexity is created by using a farm element to its optimum effect through using it as fully and intensive as possible by way of internal cycling and recycling (Werf 1993). Women farmers have evolved a system in which they would use a farm produce or element at several levels and in several functions, making waste/ intermediate product or function the input for another and ultimately cycling and recycling the end product into the farming system. In this way they would place large number of elements into a web of symbiotic and dynamic interrelationships. Larger the number of functions fulfilled by a single farm elements, the higher the degree of complexity.

The combination of diversity and complexity – high number of species and their genotypes functioning in a web of symbiotic interrelations – results in a stable farming (Werf 1993). Some of the examples of how the mountain farmers create complexity in their farming system are as follows (after Singh and Tulachan 2001):

Cereals

- ✰ Flour of the grains is consumed by humans;
- ✰ its bran is provided to milch and/or work animals;
- ✰ residues are fed to animals;
- ✰ waste produced by feed is converted to manure;
- ✰ manure is used as an input for soil fertility (nutrition for crops).

Grewia optiva (*Bhimal*) Tree

- ✰ Maintained as an important component of agroforestry system;
- ✰ leaves are used as protein-rich, tannin-free fodder for livestock;
- ✰ tender twigs and leaves are used as a shampoo for washing hair;
- ✰ fruits are used as dye;
- ✰ elixir prepared from its twigs and leaves is used as a medicine to facilitate parturition;
- ✰ its twigs and small branches are rotten to extract fiber, which is used for making ropes,

	mats, etc.; ☆ Fibre-stripped twigs and branches are use as source for light and as fuel; ☆ nutrients obtained from the tree are recycled into soil along with organic manure.
Compost	☆ Husk, straw, hay, and forest litter are added to compost; ☆ used as mulch in some crops; ☆ used for fertilizing soil; ☆ used to conserve moisture in the fields and confront drought or semi-drought conditions.
Prunus armeniaca (*Chulu*) Fruits	☆ Fruits are used for consumption at home; ☆ kernel is used for extracting good quality oil used for cooking and other purposes; ☆ fruit pulp is used for making jam, chutney, etc; ☆ cake of the kernel is used as scrub for washing face; ☆ cake is also used as a safe and effective poison to control rodents; ☆ leaves of the tree are used as livestock fodder; ☆ twigs and branches are used as fuel for cooking food; ☆ the fruit tree is a compatible alley of the agroforestry system.
Milk	☆ Used for drinking directly or through tea, coffee; ☆ converted into curd, yogurt, butter, refined butter (ghee), cheese; ☆ used for preparing a variety of sweets; ☆ sold in the market as fresh whole milk or its by-products to earn income for family.

Soil Fertility Management

An enormous number of names were applied by native agriculturists to soils. Each district rejoices in at least a dozen purely local names. Native cultivators keenly appreciate the smallest differences in the relative qualities of different soils. Names have references to external condition, frequency or decency of cultivation, its situation as regards inhabited sites, etc., its position as upland or recently formed alluvium, its occupation for pastures, fields or gardens, its external features, etc.

– ***Hume (1878)***

Marginal farmers do not regard soil as an inert, dead material, but an ecosystem, hence a part of nature's biodiversity. They follow the natural principles of nourishing the soil, so that it, in turn, nourishes the plants. They have a reverential attitude towards soil and are always cautious about its fertility status. Living soil is a precondition for a sustainable agriculture, which the marginal farmers strictly follow. In the self-contained farming systems, all they incorporate in their soil are the nutrients produced within the system. Some of farmers' strategies to manage the fertility of the soil ecosystem are given below.

Farmers transfer biomass of numerous plants in the forest ecosystems to the cropland as forest litter, or it goes after being fed to the livestock or after being used as bedding material in the animal sheds as manure. Biodiversity in the forest ecosystems will have its bearing on the soil fertility of the cropland.

Crop residues, weeds, etc., to come from cultivated land are also fed to livestock and – as manure – are recycled into the soil on which they grow. Livestock, in the process of the digestion of plant biomass in the rumen, help accelerate the process of biodegradation in nature. If the biomass were transferred to/ recycled into the cultivated land without passing it through rumen (which, functions as a wonderful micro-ecosystem), its microbial decomposition in the soil would take much longer time; and the marginal farmers cannot wait for so long.

Farmers' practice of fallowing is also meant to maintain soil fertility by containing continuous draining of soil nutrients, giving periodic chance to the land for convalescence and helping rotting of crop roots (which cannot be fed to livestock) to convert them into manure and add to soil nutrients.

Fields left fallow are used as CPRs, especially for grazing animals. While grazing, animals void dung and urine and fertilise the soil. This practice is called kraaling. In Nepal and parts of the Indian Himalayas, farmers have to pay the owners of goats and sheep against the kraaling. In-situ manuring, *i.e.* the manuring of field by tethering the animals directly in the fields is an important strategy developed by the mountain farmers over ages. During dry season, cattle (and small ruminants) graze harvested fields of finger millet left fallow for fertility regeneration and manure them directly. The draught animals also directly manure the fields, in which they perform agricultural operations by voiding their dung and urine. Results of an experiment by Powell and Ikpe (1992) show that millet yields in kraaled areas (which receive both manure plus urine) are over twice as high as yields where only manure from stall-kept animals is spread by hand. Residual benefits of urine are also observed. Yields in plots where cattle and sheep are kraaled the previous year are about 1.5 times greater than where only manure have been hand-spread.

Crop rotations, sequences, ways of exploiting complimentarity and synergy amongst various plant species and genetic resources are another efficient means of maintaining soil fertility marginal farmers have evolved over millennia of trials and errors.

Pest Management

For sustainable agriculture, increasingly uniform crops may be more vulnerable to pests and diseases.

– Joshy (2000)

Marginal farmers have developed a unique mix of crop species and their genotypes that have capability of confronting the menace of pest epidemics. Very high level of inter- and intra-species biodiversity, as a law of nature, can absorb/ repel physical and biological shocks and, thus, has high degree of resilience. Further, they have selected the landraces with natural in-built resistance mechanism.

Storage bins (the *kothar*) found in the houses of marginal farmers are often impervious to moisture and insect attack. Farmers keep walnut (*Juglans regia*) and *denkan* (*Melia azedarach*) leaves along with the seeds, which serve as natural insect repellent. Pulse seeds are treated with mustard oil before storage. This helps repel insects and preserve seed for long duration. Cow urine is also used as an insecticide.

Trees farmers grow in their agroforestry systems serve as nesting and resting places for the birds, the natural predators of insects and rodents.

Weed Management

Ecological and socio-economic values of weeds are well recognised by the marginal farmers in the mountains.

– *Shiva et al.* (2005)

Most of the weeds growing in the farmlands are no menace for the marginal farmers. Farmers rather view them as an indispensable component of the plant diversity. Growing of weeds is beneficial to them since these serve as animal feed and few of them as human food. Weeds also prevent soil from heating up or dying out excessively. They also induce a positive competition, which stimulates crop growth and reduces soil erosion.

Jakhia (*Cleome viscosa*), a highly useful plant, grows wild, but maintained as an essential complement of the finger millet-based cropping system. From a professional agriculturist's point of view, *jakhia* is a weed since it is not cultivated. From a marginal farmer's point of view, however, it is a very useful plant and its grains are found in every household.

Some of the weeds colonise on degraded slopes. They conserve soil and, in long run, allow native successional species to emerge on the site. Ecological role of such weeds is not appreciated. However, marginal farmers give value to such weeds also.

Marginal farmers manage crop diversity in such a way that weeds would hardly compete with the cultivars. When farmers find that weeds are really competing with their main crop, they remove them simply using a hand-hoe or bullock-driven weeders. They would, in most of the cases, use this as feed for their livestock. They also look for some 'weeds' for use as vegetables; *lengura* (*Diplazium* sp.), a fern, being a common example during summer season. The unusable weeds, hardly any, would be left in the field to be rotten and mixed with the soil.

However, in recent years, invasion of exotic plants, such as *Eupatorium*, *Parthenium*, and *Oxalis*, has come up as a big menace for farmers in many areas of the mountains. These exotics are replacing the indigenous plants and destroying natural biodiversity and creating an ecological imbalance in many areas, worrying the farmers. Marginal farmers are yet to cope up with this sort of menace.

Conclusion

Biodiversity is one of the most important natural resources on Earth on which humanity depends for food security, nourishment, and socio-economic and ecological development. Biodiversity has no option for survival and sustainability. Protection, conservation and further enhancement of this natural resource base are, therefore, of utmost importance. There can be no compromise with nature's biodiversity. Welfare of all living beings forming the essential contents of nature' biodiversity is the welfare of humanity. Humans are the only species amidst the nature's bounty of biodiversity, but most of this bounty suffers at the hands of this species.

Marginal mountain farmers are the beautiful examples caring for biodiversity in their traditional ways. History of Indian agriculture is replete with the examples of ever increasing level of biodiversity in the cultivated areas. The modern agriculture – or the so-called Green Revolution – however, has only served as an eclipse to the biodiversity as well as to the rich traditional wisdom caring for the biodiversity. Monoculture-based agriculture fed with overwhelming doses of chemical fertilisers and pesticides is environment-polluting and extremely vulnerable to damage.

Marginal mountain farmers are rich repositories of the traditional wisdom that helps them evolve and adopt strategies that are conservation-oriented, biodiversity-enhancing and sustainability-inducing. They have developed a balanced land use incorporating uncultivated lands under common property regime and cultivated areas. They provide biodiversity supreme importance. The agricultural traditions mountain farming communities follow are based on ecological principles and unique eco-philosophy that direct them to take care of their living soil, biodiversity-complexity, and cyclic flow patterns of the nutrients.

Traditional wisdom of the farming communities inhabiting marginal ecosystems, such as those in the Himalayan mountains, should be regarded as pristine and futuristic. This must be appreciated and given due recognition. Past is linked with future and traditions are inevitably linked with modernity. Delinking past from the future and traditions from modernity would cost humanity and the Earth. Nature's biodiversity and human wisdom have flourished together. They must be linked together, for that would be an imperative for the flourishing future.

References

Bahuguna, Sunderlal. 1987-88. Plant Wisdom to Restore our Eroded Earth. *Future*, 1987-88.

Bahuguna, Sunderlal. 1995. Chipko: The People's Movement with a Hope for the survival of Humankind. In Singh, Vir.; Sharma, R.J.; and Kumar, A. (eds) *Ecological Carnage in the Himalaya*, 85-96. Dehradun: IBD.

Dankelman, Irene and Davidson, Joan. 1988. *Women and Environment in the Third World: Alliance for the Future*. London: Earthscan.

Hume, A.O. 1878. *Agricultural Reforms in India*. (Quoted by Vijayalakshmi, K. 1994. Conserving People's Agricultural Knowledge). In Shiva, Vandana (ed) *Biodiversity Conservation: Whose Resource? Whose Knowledge?*, p.60. New Delhi: Indian National Trust for Art and Cultural Heritage (INTACH).

Jodha, N. S.; Banskota, M. and Partap, T. 1992. *Sustainable Mountain Agriculture*. New Delhi: Oxford & IBH Co.

Joshy, D. 2000. Inaugural Address delivered at the International Symposium on *Participatory Plant Breeding: Exchange of Experiences from South and South East Asia* organized by PRGA, IDRC, DFID, DDS, LI-BIRD, IPGRI, AND ICARDA, Pokhara, Nepal, May 1-5, 2000.

Partap, Tej. 1999. *Sustaining Livelihoods on Marginal Farms in the Hindu Kush-Himalayan Region: Challenges and Opportunities*. MFS Discussion Paper No. 1/ 99. Kathmandu: ICIMOD.

Powell, M. and Ikpe, F. 1992. Fertilizer Factories : Nutrient Recycling through Livestock. *ILEIA Newsletter*, 8 (3): 13-14.

Shiva, Vandana. 1991. *The Violence of the Green Revolution: Third World Agriculture, Ecology and Politics*. Penang: Third World Network.

Shiva, Vandana; Singh, Vir; Dankelman, Irene; Negi, Biju; and Singh, Shilpy. 2005. *Biodiversity, Gender and Technology in Mountain*

Agriculture: Glimpses of the Indian Central Himalayas. New Delhi: Navdanya. pp 80.

Singh, Vir. 1991. *Ecological Impact of Apple Cultivation in the Himalaya.* Dehradun: IBD.

Singh, Vir. 1998. 'Traditional Biodiversity Management Strategies in the Mountains: Farmers' Experimentation in Garhwal Himalaya.' In Singh, Vir. and Sharma, M.L. (eds) *Mountain Ecosystems: A Scenario of Unsustainability.* New Delhi: Indus, 25-66.

Singh, Vir. 2000. *Smallholder Dairy in Mixed Crop-Livestock Farming Systems in the Uttar Pradesh Himalayas, India.* Kathmandu: ICIMOD.

Singh, Vir. 2001. Biodiversity: A Key Issue for Commons' Management in the Himalayan Region. In Marothia, D. (ed.) *Institutionalising Common Pool Resources,* 519-530. New Delhi: Concept.

Singh, Vir. 2002. Appropriate Technologies for Agricultural Development in the Mountains: A Search for Sustainability-Inducing Options. Paper presented at the National Workshop on *Mountain Environment and Development: Potential and Prospect* organised by GB Pant Institute of Himalayan Environment and Development at Kosi-Katarmal, Almora, Uttaranchal, December 9-10, 2002.

Singh, Vir. 2005. Agroecosystems: Foundation for Conceptual and Practical Approaches to Transcendental and Sustainable Development of Agriculture in Mountain Areas. Paper presented at the *Traning Workshop* organized by Navdanya at Dehradun, India on September 19, 2005.

Singh, Vir. and Jardhari, Vijay. 2000. Landrace Renaissance in The Mountains: Experiences of the *Beej Bachao Andolan* in the Garhwal Himalayan Region, India. Paper presented at the *International Symposium on Participatory Plant Breeding in Asia: Exchange of Experiences,* held in Pokhara, Nepal, May 1-5, 2000.

Singh, Vir. and Tulachan, P.M. 2001. Marginal Farms and Farmers in the Himalayan Mountains: Issues and Options in Uttaranchal, India. Paper presented at the Meeting on *Issues and Options for Marginal Farmers in the Hindu Kush-Himalayan*

Region organized by International Centre for Integrated Mountain Development (ICIMOD), Kathmandu.

Skolimowski, Henryk. 1995. *Dancing Shiva in the Ecological Age: Heralding the Dawn of the Ecological Reconstruction in the 21st Century*. New Delhi: Dr. Henryk Skolimowski International Centre for Eco-Philosophy/ Indian Institute of Ecology and Environment.

Skolimowski, Henryk. 1997. Human Values and Capitalism. *World Affairs*, Vol. 1, No. 4, Oct-Dec 1997.

Tulachan, P.M. 1998. *Livestock Development in Mixed Crop Farming Systems: Lessons and Research Priorities*. Issues in Mountain Development 98/ 5. Kathmandu: ICIMOD.

Tulachan, P.M.; and Neupane, A. 1999. *Livestock in Mixed Farming Systems of the Hindu Kush-Himalayas: Trends and Sustainability*. Kathmandu: ICIMOD & FAO.

Werf, E. van der. 1993. Ecological Farming Principles. In Singh, Vir. (ed.) *Eco-crisis in the Himalaya: Causes, Consequences and Way Out*. Dehradun: IBD.